DESIGNING LIFE? GENETICS, PROCREATION AND ETHICS

Designing Life? Genetics, Procreation and Ethics

Edited by
MAUREEN JUNKER-KENNY
Trinity College Dublin

Ashgate
Aldershot • Brookfield USA • Singapore • Sydney

Published by
Ashgate Publishing Ltd
Gower House
Croft Road
Aldershot
Hants GU11 3HR
England

Ashgate Publishing Company
Old Post Road
Brookfield
Vermont 05036
USA

British Library Cataloguing in Publication Data
Designing life? : genetics, procreation and ethics. -
(Avebury series in philosophy)
1. Genetic engineering - Moral and ethical aspects
2. Bioethics
I. Junker-Kenny, Maureen
174.9'57

Library of Congress Catalog Card Number: 99-72338

ISBN 1 84014 890 X

Printed and bound by Athenaeum Press, Ltd.,
Gateshead, Tyne & Wear.

Contents

Introduction

MAUREEN JUNKER-KENNY

This book is based on a public lecture series organised by the School of Hebrew, Biblical and Theological Studies, Trinity College, Dublin, which was held in autumn 1997. It brought together representatives from different disciplines – genetics, psychology, law, and theology – in order to create a forum of debate between experts and the public. Its topic were the opportunities, dangers and choices which modern genetics and the new reproductive and cloning techniques pose to society at the threshold of the twenty-first century which has already been termed 'the century of molecular medicine.'

In the Republic of Ireland, the debate, if one can call it one, has centred on the unavailability of embryo freezing. This is indeed an issue of both practical and symbolic significance. It touches the question of the moral status and rights of the embryo over against the desire of parents for techniques which would optimise their chances for having a baby of their own at the least risk to the mother; it also marks one of the dividing lines between jurisdictions in Europe. Still, it is by no means the only issue posed by the new scientific possibilities. Other questions for public debate are the medicalisation of pregnancy and women's right to self-determination, preimplantation diagnosis and the concern of eugenics, the right (or duty?) to know or not to know about one's genetic make-up, germline interventions and the principle of consent of future generations, standards of what kind of life is still deemed to be 'liveable', and who establishes them, economic interests and hopes of minimising suffering, the protection of the most vulnerable also in relation to patients incompetent to consent, as well as human dignity and cloning. These problems of biomedical ethics can only be addressed if the social ethical, philosophical and theological background assumptions informing them are also discussed: e.g., concepts of the human person and the question of her natural determinedness or freedom, the facticity or contingency of existence and the gift of life, the rationalisation of sexuality and procreation, aims of counselling, the meaning of solidarity, control and predictability as technical goals in relation to the values of the lifeworld, and cultural standards of parent-child relationships.

At a time when conditions for moral agency seem to be undermined just when we need it most, societies have to decide on what knowledge they want, and how they will deal with it. What services, institutions, and agencies are to accompany the new possibilities of genetic diagnosis and interventions? A political debate on the implications of these exciting and ambiguous developments on our lifeworld is overdue.

Dietmar Mieth's article on 'Bioethics, Biopolitics, Theology' maps out the context in which biomedical ethics finds itself since the announcement of the cloning of the sheep known as Dolly in February 1997. He highlights the difference between two perspectives on human life: a philosophical and theological reflection on life in its givenness and contingency, and the natural science perspective that takes life as an object of research. Before discussing germline interference and cloning as problems of philosophical and theological ethics, he uncovers the politics of language which prevent the real ethical issues – e.g., how far the argument of non-instrumentalisation can be applied to embryos *in vitro* – from coming into view. This naming policy is present in distinctions such as 'non-reproductive' versus 'reproductive' cloning which veils the decision that the cloning of human embryos is permissible as long as they are not to be implanted in order to be born. He shows that what the Council of Europe's 'Convention on Human Rights and Biomedicine' (1996–98) supports is only a partial prohibition of cloning. The other test case he analyses to evaluate the extent to which rights are protected in the Convention are patients incapable of consent.

In both questions, since, the European Parliament has taken a stricter view than the compromises proposed in the Council of Europe's Convention. It has decided to rule out also 'non-reproductive' cloning and to withdraw all public funding from embryo-wasting research. In November 1998, it ruled against the provision of the Convention that medical research on patients incapable of consent could be carried out if certain conditions were met.

While Mieth understands human rights from their ethical foundation in Kant's Categorical Imperative never to treat anybody only as a means but always as an end in him- or herself, the legal expert **Neville Cox** wants to keep legal rights and ethics quite apart. In his article on 'The Failure of "Rights Talk" in the Field of Bioethics: The European Convention on Human Rights and Biomedicine' he states that it 'is entirely legitimate to be edgy about the concept of mass destruction of foetuses, of cloning, of interference of genetic lines ..., not because rights are being violated but because of the ethical implications, both religious and secular of scientists attempting to "play God"'. His detailed analysis of the same European document clearly sets out the different political bodies and legal levels in Europe as well as their consequences for implementation in national legislation. In his evaluations, he comes to rather different conclusions than Mieth. 'To afford such a gigantic legal status to the embryo, an entity whose moral status, far from being above moral controversy actually generates it, is certainly dangerous and arguably wrong.' He views criticism of the Convention on account of its lack of respect for the integrity of incompetent patients as 'a rather melodramatic response'. While Mieth fears that the role which biopolitics envisages for ethics is that of an enforcement strategy, Cox is concerned to distinguish morality from prejudice if morality is to be enforced by the law.

While the first two papers from the neighbouring fields of ethics and law thus open up a wide arena of controversial debate, the following two articles represent perspectives from the natural sciences.

In his contribution entitled, 'The Machine in Man', **David McConnell** explains the methods of modern genetics which are used in the key technology of genetic engineering. He then confronts the problem of the determinist framework of genetic enquiry and identifies 'two glimmers of hope for non-determinism': the physical processes of the brain and the human faculty for language. Whether the philosophical or metaphysical perspective on human life will be made redundant by the scientific view the more it finds out about the beginning of life and the neurological mechanics of brain activity is a question which remains open. He concludes with the hope for renewed attempts at cooperation between the natural sciences and the human sciences including philosophy and theology which was realised in an exemplary way in the 1940s and 1950s, by the natural scientist Julian Huxley and the biologist of evolution and theologian Teilhard de Chardin.

In his paper, 'Designing Life: The Role of Inheritance', **Dan Bradley** describes the significance of genetic and environmental factors in shaping personal traits. He emphasizes the difference between statistical results and individual prognoses and takes care to indicate precisely what kind of claims can be made from the angle of genetics. In a historical backdrop, he shows how the assumptions of the eugenics movement are questionable both at an ethical and at a scientific level. The article concludes with the ethical dilemmas posed by the new possibilities of genetic diagnosis.

Sheila Greene's contribution, entitled 'What Makes a Person a Person? The Limits and Limitations of Genetics', challenges the reductionist assumptions behind some approaches in genetics, but equally in her own discipline, psychology. Against the objectification and depersonalisation of human beings evident, e.g., in their description as 'robots' by R. Dawkins, she defends the crucial concept of human agency. She questions the 'allegedly more privileged perspective' of reductive explanations of human behaviour as programmed by their genes. In order to understand human praxis in its specificity and complexity, the capacity for rationality, accountability, reflexivity and intersubjective recognition of persons in their uniqueness has to be taken seriously. Her warning that the new genetic knowledge equally includes 'great opportunities for evil and recklessness' is already borne out e.g., by recent media reports on attempts to construct an 'ethno-bomb' that seeks out and destroys the genes of peoples designated as 'enemies'.

The three final contributions from women theologians engaged in the interdisciplinary debates of biomedical ethics take up this perspective of the inherent limitations of human intentions and agency. Besides the God-given inventiveness and freedom of the human person, theological anthropology equally highlights human temptability, fallibility and finitude.

Helga Kuhlmann's article on 'Crisis Pregnancies in the Age of Human Genetic Diagnosis. Women's Right to Self-determined Pregnancies and the Right of the "Other"' gives an overview of positions held in the feminist debate on their reproductive rights, including the voice of disabled women with regard to freedom of choice on whether to continue a pregnancy. She shows how the evaluation of embryonic life by the state and the churches has changed historically. In conclusion, she proposes ways in which the churches and Christian theology can help to create conditions for greater self-determination of women in crisis pregnancies.

In her paper, 'Ethical Reflections on Genetic Counselling in Prenatal Diagnosis', **Hille Haker** continues the debate on who is a person by supplementing the approach of founding human dignity on agency with the phenomenological philosophical tradition. She suggests a more inclusive view of human personhood which comprises the conditions of temporality, bodily constitution and intersubjective recognition. Personhood and the protection of human rights cannot therefore only be attributed to fully-fledged autonomous rational beings. In her exploration of the ethical presuppositions of genetic counselling she maps out the levels of ethical reflection to be addressed in counselling which does not lose its nondirective character by taking the conflict of rights between the parents and the foetus seriously.

While Haker is dealing with the question of the rights of the human foetus in prenatal genetic counselling, my own essay on 'Embryos *in vitro*, Personhood, and Rights' considers argumentations for and against the personhood or person status of the human embryo. I then explore the impact of cultural changes affecting the lifeworld through individualisation and market imperatives on reproductive and diagnostic choices. The concluding part deals with the question of what style of argumentation a theological ethics engaged in interdisciplinary debate should develop to promote insights into the ethical values fundamental to pluralist societies with diverse philosophical, ethical, and religious traditions.

My thanks are due to all the contributors, to David McConnell for his support in staging the lecture series, and to Dr Fergus O'Ferrall of the Adelaide Hospital Society, the financial support of which made it possible to invite speakers from abroad. Sincere thanks also to Ms Jane Welch, secretary of the School of Hebrew, Biblical and Theological Studies, for all the technical organisation that goes into a public lecture series, and to my husband, Peter P. Kenny, for designing the title.

1 Bioethics, Biopolitics, Theology

DIETMAR MIETH

Introduction

In a famous lecture, given in Chicago 1989 to the Evangelical Lutheran Church in America on 'The Institutional Construction of a New Fetish: Human Life', Ivan Illich started with the following thesis:

> 'Human Life' is a recent social construct, something which we now take so much for granted that we dare not seriously question it. I propose that the Church exorcise references to the new substantive life from its own discourse.

For Illich, the notion 'life', which is so essential for modern ecological, medical and ethical discourses, is in the Western tradition 'the result of a perversion of the Christian message'. The notion 'life' under contemporary conditions does not belong to the world of sacred and contemplative feelings. It is a word which belongs to the field of modern management, to the language of planning so called 'human resources'. The context of epistemic sentimentality about an unquestioned progress makes of 'better life' a new fetish. This created the struggle between the two contradictory options between 'pro life' and 'pro choice'. But this discourse is forgetting the connotation of non-objectivation, which was primarily given with the gospel of life. The gospel of life is personal 'I am life', from Moses to Jesus. The modern notion is a 'value', a good, which must be preserved or destroyed according to social choices.

Illich presents for us five observations on the history of life, which we should not forget. 'First, life, as a substantive notion, makes its appearance around 1801.' Instead of the religious and philosophical tradition of psyche, bios and zoe, the term 'biology' means 'a science of life' (Jean-Baptiste Lamarck). 'Life' is from now on a construct of organic phenomena like reproduction, genetic development and so on.

'Second, the loss of contingency, the death of nature and the appearance of life are but distinct aspects of the same consciousness.' The loss of contingency here is the loss of dependent and actual connection with the breath of creation. The mechanistic model replaces the creative-processual model. 'Third, the ideology of possessive individualism has shaped the way life could be talked about as a property.' It can easily be demonstrated in the patenting-life discourse that life is being discussed in its 'elements' not as a matter of 'discovery' but as a matter of human 'invention' even if it is identical to its natural state. On the other hand, the instrumentalisation of human life, which we will demonstrate in the options concerning cloning, is clearly a result of this 'possessive individualism'.

'Fourth, the fetitious nature of life appears with special poignancy in ecological discourse.' 'To think' life as a system of correlations between living forms and their habitat is a reduction of imagination and a surrendering of life to all kinds of empirical and also virtual objectivation.

Fifth, the pop-science fetish *a* life tends to void the legal notion of person. The *distinction* between 'human life' and 'human person' created the notion of a 'human non-person', which is not a member of the so-called 'moral community'. 'The new discipline of bioethics', as Illich concludes, 'mediates between pop-science and law by creating the semblance of a moral discourse that roots personhood in the qualitative evaluation of the fetish, life.' I refer to these critical observations of Illich as an important perspective of so called 'bioethical' questions, because we are seldom aware that, before beginning an ethical discourse, we have to 'exorcise' a language and a language politics which does not allow a real ethical approach, only the semblance of it. 'Bioethics' is an invention of the scientific language of 'biology' and its derivations. This paradigm is dominating ethical, legal and social responsibility. This can be demonstrated by an example: the so-called 'ELSA' group, which is responsible for the research projects of the European community concerning the 'ethical, legal and social aspects' of biotechnology, in 1997 had 14 members. Of these 12 were biologists, one was a philosopher and one a lawyer. It is clear that there is also an interdisciplinary approach to ethics and that interdisciplinarity makes sense in ethics concerning the sciences and new technologies. Nevertheless, it is not incorrect to suggest that the paradigm of the life fetish as a fetish of a scientific paradigm is strongly present in this kind of discourse. Therefore, as I try to demonstrate in my concluding theological proposals, the exorcism forced by ideas such as contingency is an important precondition for an 'autonomous' ethical approach in a theological context.

Bio-political Experiences: The Politics of Language and the Argument of Human Dignity

Ethical consultation of politics is meant to serve the purpose of developing legal frameworks which then determine the conditions under which research and the application of research can be carried out. This consultation on politics occurs on different levels. I will first clarify which levels exist and in which level I am involved. The first level which has to be considered is the level of UNESCO.

At the level of UNESCO, from 1994–97 the International Bioethics Commission (IBC) prepared a text which was accepted by UNESCO in November, 1997. This text is a declaration aiming to protect the human genome.

The next level is that of the European Council. The ethical consultation group of the European Council is called the Steering Committee for Bioethics. It elaborated the drafts of the European Convention on Human Rights on Bioethics. At the level of UNESCO we are dealing with a declaration, that is, an appeal which bears a moral obligation. At the level of the European Council, by contrast, a contractual framework has been developed which becomes valid as soon as the individual member-states have not only signed but also ratified it.

A new level is represented by the European Union of 15 countries which, in the transition to becoming a federation, has a parliament and a government. The government is called the European Commission. It is assisted by an advisory group called the Group of Advisors on Ethical Implications of Biotechnology. From 1994–97, I belonged to this nine-member group and since 1 January 1998 to the succeeding 12-member group (European Group on Ethics). On the European Commission's instructions this advisory group has delivered several statements which became important, for example, with regard to the question of patenting, the question of cloning, or the question of food labelling.

Besides this, there arc advisory groups at the level of the World Health Organization. There, too, a statement on human cloning was worked out. Of course, there are also advisory groups on the national level. All of them have in one way or another commented on the theme of cloning. Thus, this example is illustrative.

The Dolly sensation is well known since February of the year 1997. The common term for this type of cloning technique is 'stem cell' cloning. We do not know to what extent the experiment is repeatable; we also do not know to what extent it is applicable to humans. Yet the suggestion has already been made to apply this stem cell cloning technique to humans. Thus, about 22 famous scientists throughout the world have advocated the use of this technique on humans. The American biophysicist Seed attracted attention with his suggestion that this type of cloning technique could be applied to help infertile men to have children. The technique to clone germ cells caused great public excitement because it conveyed the idea of the copying of existing human beings. But it is simply impossible to talk about a total genetic identity, if for no other reason than that a whole series of other factors from the environment influence human development. Already from the genetic viewpoint it would be wrong to assume that all 'copies' that have been cloned look the same among animals, for example. They don't even necessarily have their spots in the same place, or the same physiological shape.

Already in 1993 attempts were made in Pittsburgh to double embryos (by splitting). Of course the embryos were not implanted and no development of twins was brought about this way. As soon as we are talking about cloning today, embryo-splitting is always included, especially with regard to *in vitro* cloning techniques. We have to differentiate between cloning techniques with the intent of *in vivo* development, that is of copying humans, and cloning techniques which are limited to the *in vitro* phase, to the embryos before implantation. By transforming them into cell-cultures particular purposes are pursued, for example in connection with therapies to prevent immune-rejection in the case of transplantation, and also in connection with the development of early forms of human organs.

At this point we are still talking about science fiction. However, we are dealing with interesting situations in which a society, with the help of ethical reflection, can declare its position before the technique as such exists. The various advisory groups were commissioned to present experts' opinions in order to reach ethical standards: the Clinton commission in the US, the GAEIB, the already mentioned advisory group to the European Commission which completed its 'Opinion' on 20 May 1997, in other words, under great time pressure, and the advisory group to the German Minister of Science, in which the presidents of various research organisations participated. With regard to the cloning of human beings, a distinction was introduced in the EU and the US Commission advisory groups between so-called 'reproductive' and 'nonreproductive' cloning. I have already mentioned that we have to differentiate between the *in vivo* and *in vitro* situations. Our language is not capable of grasping all new phenomena immediately, thus any concept one can choose has the potential for being misunderstood and this may also be applied to the differentiation between 'reproductive' and 'non-reproductive' cloning. 'Reproductive' means that the clone is implanted and a human being develops from it. The *in vitro* situation is assumed. *In vitro* does not yet determine whether an early embryo, which has been manipulated accordingly, will or will not be implanted. And if it is not implanted, but rather experiments are carried out on it with particular long-term therapeutic goals, then this counts as 'non-reproductive' cloning.

The advisory group of the EU came to the conclusion that 'reproductive' cloning must be forbidden. The Clinton Commission, using the same terminology, was more liberal only insofar as it demanded a moratorium 'at present'. The formula 'at the present time and the present social context' did not enter into the conclusion of the European advisory group because I was opposed to it. In my opinion, this limitation assumes that this prohibition will be revised. It is self-evident that every rule which we formulate under present circumstances can be revised in the future. Yet whoever wants to bring this to expression has ulterior motives which have to be put on the table. Thus it can be said that the Clinton Commission declared itself in favour of a moratorium whereas the European advisory group was in favour of a strict ban, but only in the so-called 'reproductive' area — I will come to the reasons behind the strict ban later. In the 'non-reproductive' area, that is, *in vitro* cultivation of cloned embryos in cell cultures, the EU advisory group stated that in those countries which allow experimentation on embryos (such as, e.g., Belgium or Great Britain) *in vitro* cloning should not also be forbidden on the condition that one is dealing with high priority therapeutic purposes, as well as on the condition that a licensing body, that is, an ethics commission, is consulted and finally, on the condition that the manipulated embryos will not be implanted and become independent human beings.

Even so, I did not sign the declaration of the EU advisory group because I am of the opinion that the differentiation of 'reproductive' and 'nonreproductive' cloning was not decided on a *factual* basis, rather, it had the intention to thereby reach a differentiation in the treatment of 'embryos' and 'humans'. Also I did not sign because in a 1994 document I found a statement by the ministerial council of the Council of Europe regarding tissue-banks in which 'reproductive' and 'nonreproductive' were defined as to how they are to be understood in this arena. There, egg cells, sperm cells and embryos were considered to be 'reproductive'. Their use as tissue was to be forbidden exactly because they are 'reproductive', that is because from these human life could develop. The expression 'reproductive' thus was intentionally narrowed clown to implanted embryos by the advisory group in the interest of practicability. This I call the *politics of language*. This politics was not without precedence.

This difference between 'reproductive' and 'non-reproductive' has previously been introduced by an American advisory group on *in vitro* fertilisation.

Non-reproductive cloning today is called 'therapeutic cloning': this is also a politics of language. The concept of therapy or the concept of human health plays a very central role in the politics of language, therefore I would like to elucidate this with an example from the so-called Bioethics Convention of the European Council. Article 18.2 of the convention states that the production 'of embryos for research purposes' is forbidden. Article 12 implicitly deals with research on embryos, although explicitly it is concerned with the possibility of sex selection. There it is stated that this is possible only if it serves health purposes. As an example of health purposes, sex selection in order to prevent hereditary disease is mentioned since a hereditary disease can be connected to a particular gender. The text does not clarify about which method it refers to: abortion after prenatal diagnosis, embryo selection before implantation, possibly also sex selection through the centrifuging of semen. At the same time, Article 12 states that these health purposes also include 'research on health purposes'. Of course, one wonders at this point what is meant by the prohibition of embryo production for research purposes in Article 18.2? Are research purposes for health purposes – which elsewhere are always allowed – excluded? If this were the case, then the research purposes of article 18.2 would mean little, because one could declare any research purpose in this area to have a therapeutic purpose.

With this example I want to make clear that by means of such politics of language the Bioethics Convention simply puts certain problems off without solving them. This is a problem which we could illustrate with further examples. The Bioethics Convention as well as its supplementary report on cloning which was signed in January 1998 leaves the establishment of what is a human being to the nation-states. In an interview, the president of our advisory group, Noelle Lenoir, explained 'when human life begins is determined by the nation-states'. One can imagine that, in the attempt to reach a valid European consensus, the openness of this question results in any consensus remaining unclear in the specific case. According to Noelle Lenoir, Germans consider an early embryo to be a human being, other countries only recognise human beings after birth. In the Human Rights Convention on Biomedicine it is not determined what a 'human being' is.

This has particular significance for the cloning of human beings. It concerns the supplementary report (January 1998) which had been so highly praised in the European press. In this supplementary report it is stated that it is forbidden to create a human being with the intention of making him or her identical to an already living or deceased person. Someone reading this without bias assumes that this means that any human cloning is forbidden. Yet things are not so simple. The misleading political language is intensified even more by the fact that in the second paragraph it is stated that this prohibition is so strict that there are no exceptions. However when consulting the official explanation things look different. This explanation was not included by the Federal Ministry of Justice in Germany when it first published this text, which I also find very interesting.

In this explanation it is stated that one has to distinguish between three levels of cloning. First, the cloning of cells in general is morally not problematic. I am also of this opinion. Cloning is not morally problematic when dealing with a living entity without an independent destiny. Second, *in vitro* cloning, here 'embryo cells' are carefully looked at. But behind 'embryo cells' in the totipotential state embryos are hidden. Third, the cloning of 'human beings'. This means that the cloning of 'human beings' is being distinguished from the cloning of embryo cells, and then it is expressedly indicated that this supplementary report applies only to the cloning of 'human beings'. Thus, I come back to my discussion of the politics of language: although the expression 'nonreproductive cloning' is not used here, it is clear that the technical differentiation which is meant by it – *in vivo* cloning no, and *in vitro*, yes – was adopted by the Human Rights Convention on Biomedicine. Now some interpreters of the Human Rights Convention on Biomedicine claim that this is not that serious since it is stated in the aforementioned Article 18.2 that one may not produce embryos for research purposes, and this excludes the possibility of using germ cells to create embryos. Yet I have already noted that the concept of 'research purposes' is not yet clear. Is research for health purposes really excluded by this? In the future the courts will seek to clarify this since at present it is not clear from the text. Thus, one cannot assume that there is a total prohibition of human cloning in the Human Rights Convention on Biomedicine and its supplementary report. Such a prohibition does not exist at all in the framework of the UNESCO declaration, since in the UNESCO declaration only 'reproductive' cloning is forbidden. The UNESCO declaration adopted this expression from the EU-advisory group.

The importance of this 'politics of language' becomes clear when taking the fact into account that the representative of the Vatican to the consultation of the UNESCO declaration supposedly supported the prohibition of 'reproductive' cloning strongly without realising that this was only a partial prohibition. This whole linguistic game cannot be immediately recognised by theologians since they do not constantly have to deal with these matters.

This distinction also plays an important role in the cloning prohibition which is demanded by the World Health Organization. Their almost identical words forbid the creation of a human being as a copy of another, and this always means '*in vivo*'. Thus internationally, there is only one place so far where one may be able to speak of a total prohibition of human cloning. The position of the advisory group of the European Union, which I refused to sign, led to a controversy in the EU Parliament. The Parliament strictly outlined its position. In a decision on human rights of 15 January 1998 it stated:

> concerned about the announcement of an American researcher that he intends to clone human beings, with the understanding that human cloning is defined as the creation of human embryos with the same genotype as an other distinct living or already deceased human being in his or her total development, beginning with conception, without difference as to the method used, this parliament again confirms that each individual has the right to a genetic identity and that the cloning of human beings must be forbidden without any exception.

This means that the European Parliament, aware of this politics of language, had done everything possible to assert a total ban on cloning at least at the level of the EU, and this is not without significance since the European Union, in accordance with the Parliament, administers research funds. Our EU advisory group delivered an ethical framework to the EU at the end of 1997 which is meant to serve the purpose of awarding research funds according to ethical principles and an ethical evaluation of the research programmes. Certainly this ethical evaluation is broadly defined in order to be pluralistic enough, but the European Parliament will determine the principles for the authorisation of research funds in the European Union also in the future.

I would now like to deal with the reasons supporting the argument. At the level of UNESCO, in the Clinton Commission despite its weaker objection, and also in the European Union at the level of the European Council in the Human Rights Convention on Biomedicine there is obviously overall agreement that *in vivo* cloning should not be allowed, that is, no cloning that results in the creation of human beings as individual copies. Yet this type of cloning is nonetheless being aimed for. There is, for example, the suggestion of circumventing male infertility by using the man's germ cell together with the woman's ovum and in this way to make children possible for a married couple. Seed advocated this as an extension of existing reproductive medicine. Another option is that parents who have lost a child through a traffic accident and who for a 'replacement' can use one cell of this child in order to clone another one which is supposed to bring the parents the joy of repetition. This option is already problematic for purely technical reasons, but also for psychological ones. Such intentions were seriously proposed as ethically irrefutable by ethicists on the commissions whom I know. A third example originates from P. Kitcher, who wrote a philosophical contribution on methods of genetic engineering. A child has kidney problems, for example, and the parents breed a second child in order to have a spare kidney which will not bring with it the problem of possibly being rejected. Stem cell cloning could be a good method to achieve such similarity between the organs.

The reason for forbidding such options has been the same at every level of political consultation. The ethical principle is the 'non-instrumentalisation' of the person. The principle of human dignity according to Kant is: act in such a way that you never treat the other just as a means to an end, but rather always also as an end in himself or herself. The person with whom we are interacting should always also be an end in himself or herself. This is not a principle forbidding every sort of instrumentalisation, but rather, a principle forbidding the instrumentalisation of the existence and the essence of the whole person, because the person's end-in-him/herselfness is to be preserved. This means that a person who has to see himself or herself as the copy of another upon the request of a third party would be instrumentalised as such by the request of the third party, and this would specifically apply to all three examples which I presented. The request of the third party, in this case the wish of the parents, and this is important, does not only relate to the existence of a person. It is not just a matter of a person coming into existence in the sense of wishing for a child of one's own, it is a matter of design; it is a matter of the specificity of this person. Or it is an intervention in the essence of the future person in such a way that he or she serves purposes which were externally imposed. At first glance this argumentation seems very convincing, especially in our tradition. But one must not hide the fact that there are strong counter-arguments. These are presented by P. Kitcher and D. Beyleveld. They recognise the Kantian principle: act in such a way that you never regard the other as only a means to an end, but also always as an end in himself or herself. Thus, we are not dealing with an argument against Kant. The claim is made, however, that we are not dealing with total instrumentalisation if a second child were created for purposes of transplantation of organ because this child would not be restricted in its moral capacity, which is, after all, what really goes to make up human dignity. This second child, after all, would be free, have the capacity to act, and moral capacity, and as long as the capacity to act, that is, the moral capacity, of a future being is not restricted, one can only speak of an instrumentalisation that is consistent with human dignity.

I reject this. It is problematic to simultaneously treat anthropological qualities which are necessary for moral capacity as anthropological qualities which are necessary in order to respect the other. I consider this to be an 'ethicistic' fallacy (see D. Mieth, *Moral und Erfahrung II*, 1998) when one uses the anthropological reductions – which we need in order to reach ethical judgments and to ground morality – directly for the specific *practical* applications. The capacity to act, moral capacity, freedom, etc. are such reductions because this is not how the human being can be conceived in all stages of life and in their complexity. I am of the opinion that when we speak of human beings being the object of our recognition and regard, we may not transfer the normative reasons for *justifying this* recognition directly to the specific *practical* application, because otherwise it will always be persons who are somehow defective or not yet fully developed in their capacity for freedom, action and morality who will easily suffer as a result. Interchanging moral reasoning and the anthropologically determined application of morality to me seems to be wrong.

Thus, I hold on to the principle of non-instrumentalisation and am of the opinion that it is right to condemn *in vivo* human cloning. Worldwide, in 1997, we had a once-in-a-lifetime chance to reach such a prohibition due to the general alarm.

This prohibition, however, as I tried to show, is not total. It is an *in vivo* prohibition. It does not include the *in vitro* prohibition, and with that the question is posed to what extent the argument of non-instrumentalisation can be applied *in vitro*. As far as political consultation is concerned, the supplementary report on the protection of embryos for the Human Rights Convention on Biomedicine of the European Council must still be prepared. The supplementary report on cloning also refers to this missing supplementary report on the protection of embryos. Particularly the discussion in the German-speaking countries about the Human Rights Convention on Biomedicine which is, after all, not only concerned with this problem, depends on it.

In conclusion I want to take a brief look at so-called germline cell therapy. Here, the word 'therapy' is also political language. We do not yet have this therapy, we do not even have somatic gene therapy. We have trials in the area of somatic gene transfer, some of which are promising, but in the sense of a clinically applicable, repeatable, approved 'therapy' we have none of this yet, thus the word 'therapy' is one of the ideologically significant words which appear in this context particularly in the phraseology of 'therapeutic' *in vitro* cloning.

The UNESCO declaration addresses also germline cell manipulation only indirectly. It also contains one article which awards compensation to persons who suffer harm because of a genetic intervention. This was obviously intended to be a warning to researchers against doing something which could possibly result in someone in the third generation having a claim to compensation. Yet this is a very weak way in which to bring up this topic, and it is contradicted by Article 13, which states the right of all to the 'benefits' of genetic engineering.

This is also a well known 'trick' in ethical debate: to say that all persons should have equal access to the advantages offered by the new technologies. This principle of *access* easily turns into a principle of *coercion,* when, for example, it is suggested to developing nations that they should promote precisely this technology so that they, too, will have 'equal access'. Here we are also dealing with a question of cultural autonomy, to which extent these countries are able to reach conclusions *themselves* as to *what* they consider important, to *which* technologies they want to have *what kind of* access. While Article 13 of the Human Rights Convention on Biomedicine does not explicitly mention germline intervention, still it mentions it indirectly. It is forbidden to intervene with the *goal* of passing certain characteristics on to the next generation. Clearly, this is intended to be a restriction on goals. Yet there can very well be manipulations in the sense of genetic transfers of which the primary goal is *not* the transmission of the results of the manipulations to future generations. One can imagine many cases which, while they are not aiming for this *goal*, nonetheless result in having a side-effect on the next generation. Thus, the Biomedicine Convention contains only a soft prohibition.

Now one wonders, are there ethical arguments in favour of a stronger prohibition? While I do not have any categorical arguments against germline cell therapy, I do have arguments that are very strong even though they are not completely independent of the circumstances. The first argument is the uncertainty argument. It deals not only with the risk entailed but with the *uncertainty* of the risk we are willing to accept since we do not have previous knowledge about the interaction between genes and the environment and since we can only gather experience by already doing what is supposed to be the object of evaluation. This is a problem with all technology risk assessments, since such assessments always require to run the procedure of which one is first supposed to assess the risk. Thus technology risk assessment is an indirect form of technology promotion. This is also a question of the politics of language. However, what I am concerned about here is the argument with uncertainty.

The second argument is the irreversibility of intervention in the germline.

One of the important controllers of research for the European Union is J. Quintanilha – he leads the aforementioned group of 12 biologists, one philosopher and one lawyer which examines the ethical, legal, and social questions raised by research projects (ELSA) and which awards research projects in this area. The president is obviously also a biologist, which means that the political question is raised: to what extent do the biologists determine which ethical, legal and social problems are addressed by the group rather than the other way around, the problems raised by biology determine what is clarified by ethicists, social researchers and lawyers. This is the situation in the European Union. Quintanilha is of the opinion that if we had the technical ability to develop germline cell therapies, then the result would also be reversible. He says that this technique is so difficult that if we were to achieve it, we would also be in a position to be able to reverse it.

The third reason which is raised again and again and which has to be discussed states that the generation that results from this manipulation cannot consent to it. Thus, there is a lack of 'informed consent'. It is possible, due to the multifaceted functions of single genes, that a genetic intervention will not only result in advantages but also in certain disadvantages. Since ethicists in the whole area of genetic technologies have reached consensus that individuals in a democratic society should have the right to decide about the application of advantages and disadvantages, this concern carries weight.

These three arguments form a sort of cable made of different cords, a 'convergence argument' which judges germline cell therapy to be ethically negative. However, I emphasise: the UNESCO declaration does not contain a prohibition and the prohibition is only very weakly expressed in the European Convention. The leader of ELSA, Quintanilha, is of the opinion that we strictly determine our lives socially, that we have a set design for our social life, that we raise children in particular ways, that we nourish ourselves in particular ways, that we have a particular culture, that we are dominated by this whole system of science, technology, and economy. We already intervene significantly in that which makes us human, we already determine human life to a large extent. Thus, he offered this *social* design of our lives as the reason why we could also change the *natural* design through cloning and through negative selection. This type of argument appears just as frequently in the bio-political ethics strategy: why should we refrain from doing that which exists in line with options we no longer question but rather which we unquestioningly pursue.

The answer to this is clear: we have good ethical reasons to restrict determinism in social design. This is why we strive for ethical autonomy in two senses: free self-determination and free self-obligation, the unity of 'autos' and 'nomos'. Thus, we in no way resign ourselves 'morally' to actual determinism. Ethics is doubtful about social determinism. Therefore the objection to a natural design as heteronomous of the right life is justified.

We have seen that biopolitics gives ethics strategic significance yet at the same time it reduces it to an enforcement strategy. The relationship between biopolitics and ethics is in fact dominated by biopolitics in the ethics advisory groups. This is not good. Only when moral judgments are *pre*-political – which does not mean they must be *un*political – will political ethics again be ethics, *before* politics accommodates ethics to itself.

Theological Arguments on the Ethics of Human Genetics

If one accepts the concept of an autonomous morality in the context of Christian ethics, as do many German-speaking moral theologians (J. Fuchs, A. Auer, F. Böckle, et al.) then this does not yet mean a decision in favour of a particular philosophical way of thinking in ethics, but only the insight that the basis for judgments is achieved by philosophical means. Assigned to theology is the context of the discovery of ethical values and obligations rather than primarily the context of justification. Furthermore, theology can contribute significantly to ethical motivation, that is, to the motivation for ethics. Finally, theological treatments of ethics – one just has to look at the doctrine of grace and justification – have a very specific function. Theology is relevant to the discovery of ethical problems, to moral sensitising, to moral motivation and to the relativising of moral judgment on persons. In contrast to Kant the epitome of being human is not moral capacity but rather the need for salvation, the 'feeling of absolute dependency', to complement Kant's anthropology with Schleiermacher's. Happiness, freedom, and God are not philosophical postulates, but rather specific religious experiences.

Given this, it is not to be expected that theology will replace, circumvent, or change the ethical arguments in the biomedical and biotechnological debate in general or in the debate about human genetics in particular. This is to be illustrated by an example. In the debate about the protection of embryos, the documents of the Catholic church do emphasise that from the beginning humans are the image of God, and thus carriers of dignity before God, that is, of personhood (person and human dignity are interchangeable in church documents). However, this doctrine does not spare us from the need for philosophical reflection in which we must first explain which living entity is meant when speaking of God's image. We should not attempt to replace this reflection neither with false biblicism or with false positivism of doctrine – the first is the Protestant, the second the Catholic temptation. This could cause abstruse speculations to come into being. In line with biblicism, for example, the words of the psalmist, 'You drew me from my mother's womb, since I was in my mother's lap you have been my God' (22: 10, 11) would mean that being the image of God should be understood as beginning at birth. Or the word of Psalm 51: 5, 'my mother conceived me in sin', could be understood to mean that personhood begins at conception. The bible cannot be applied to answer questions which did not exist in biblical times. It is more likely that the way of discourse can serve as a model for solving controversial problems.

A positivism that regards the fertilised egg as a person only for reasons of caution (so-called tutiorism) would not help either. It could, however, be reinforced by philosophical arguments about the continuity of the human from the very beginning. But every philosophical argument remains vulnerable and hypothetical, it is never definitive.

If ethical rightness or rather its anthropological conditioning, does not belong to the theological deposit (except possibly in that tautological form which prohibits murder, where murder is by definition unethical), then the theological argument can 'only' address ethical insights contextually. It is ethically relevant without making philosophical methods irrelevant. With that, the significance of theology is in no way played down or diminished. The significance of such contextual arguments can be that they change the whole scenery of ethics and, with that, they can significantly influence the relative importance of ethical arguments. In the following paragraphs I wish to illustrate this.

Asked whether there is a relevant theological argument against the cloning of human beings I have indicated, without neglecting the previous explanations, that the diversity of humanity is a commandment of creation. This is Ellen van Wolde's conclusion in the examination of Genesis 11. The many languages imposed by God are not an answer to human hubris, since towers that reached up to Heaven were in no way interpreted as storming Heaven. (This has been better known among orientalists than among preachers!) It was the urban centralisation of humanity, their inclination to uniformity in culture (and religion?) that displeased God. Therefore he reminded them of the commandment of creation to fill the whole earth, and he dispersed humanity through the confusion of tongues. Something similar was repeated at Pentecost: the gospel, communicable despite the many languages, was spread throughout the world. Of course, diversity, not repetition of the same, can only be applied as a theological motif if we have good *ethical* arguments in its favour, and we have these as I tried to show above. Just as the argument about the person being the image of God and the argument about dignity are only inspirational in ethics, but are not substitutes, is the thought about diversity also a theological stimulant but not a substitute for arguments.

The stimulating effects of such theological impulses transcend the problem itself. This can be shown by a further theologoumenon of ethical relevance: finitude, that is to say, contingency. Obviously, contingency is also a philosophical concept. Of course, this concept is rather removed from theology when it is combined with 'coincidence,' such as in Richard Rorty's post-modern interpretation. In classical philosophy, on the other hand, contingency rather has the meaning: the dependence of being, being time dependent, finite in the sense of ending at death, capable of mistakes, the imperfection of all that is human. This is exactly what is meant by the theological theorem of the creatureliness of humanity – of not being like God. Seen theologically as well as philosophically – both ways are united in Pascal, for example – trying to transcend their own finitude humanity would gamble away the quality which makes up its humanity. Theology expresses this in a narrative form in the story of the Fall. Philosophy draws attention to the fact that it is precisely in the corporeality of being human where both fulfilment and want or deprivation are located (e.g. according to Merleau-Ponty), that finitude is the form of the possibility of happiness.

Finiteness could be a concept of convergence between theology and philosophy. It is relevant in ethics because it interprets the following as missing the point of what it means to be human; the expectation of humanity's perfectibility, of the principal feasibility of achieving anything, and of being able to still correct all problems that are created anew by problem-solving. This interpretation considerably changes the context of the biomedical and biotechnical debate. It is, for example, a stimulans for the ethical rule of problem-solving: one should not solve problems in such a way that the problems which arise through the solution are greater than the problems with are solved. I do not believe that the paradigm of progress in modern biotechnology, including human genetics, keeps to this contextual insight. In bio-politics I have observed that in this regard continental Europeans, even if they are agnostics, can be convinced of this insight more easily than North Americans, even if they consider themselves to be Christians. How is this possible?

In conclusion, after the theological arguments of being made in the image of God, of creation's commandment of diversity, and of finitude I want to introduce a fourth – which I often painfully experience as missing from the context of reflection: the vulnerability of humanity. Analogously to the theological 'option for the poor' in our context (biotechnology, human genetics) it is possible to formulate an option for the priority of vulnerable persons. Here one easily runs into a lack of understanding in the secular context. Certainly, the average pragmatist who does ethics in political advisory groups will acknowledge that one must respect disabled persons and incompetent persons, even (more or less) foetuses and embryos. But he or she will face a theological theory such as 'the prophecy of disabled persons' with little understanding. And yet, still it is easy to explain to him what it means: an evidently disabled person reminds us that we are all disabled in some way. It reminds us of our finiteness.

In the practice of finding ethical judgments in an accelerated world of suspended doubt one often meets the readiness to represent persons who are not able to realise their interests and opinions by themselves that is, who cannot express themselves, by extrapolating their presumed intention via analogous deduction. The fiction of 'informed consent' is stretched to include them. Curiously, exactly in this way they tend to be instrumentalised. When, for example, the European Council's Convention on Human Rights and Biomedicine tried to formulate the conditions under which research for the benefit of others can be carried out on those incapable of consent, it forgot to secure in the document the ability to withdraw proxy consent, as was discovered by my Protestant colleague from Vienna, Ulrich Körtner. A whole mindset can be recognised in this forgetfulness!

There is an aura of vulnerability around persons who cannot represent themselves, the shimmer of the butterfly that may not be touched. Forgive the poetic expressiveness, but I want to illustrate that we are dealing here with the need for greater sensitivity. Sensitivity can be easily declaimed. We then carry the burden of proof that it is not present. Our time lives with the ideal of the person as young and strong, an ideal for which Paul already admonished the Corinthians. It is not just a matter of developing a culture of perception and feeling (although also of this). It is a matter of the role of taboos which have grown out of historical experiences, for example the Nurnberg code which simply forbids experimentation on those not capable of self-determination. Taboos can only be ethically justified if they are understood as cautionary fences. They do not belong in an ethical justification but in its context which modifies the importance of the justification. Or stated differently: situations of vulnerability demand caution. I keep asking myself why the Bioethics Convention of the European Council did not allow advocates for the affected groups themselves to formulate the norms regulating non-therapeutic research on those incapable of granting consent. Here, a sense of superiority of the experts is expressed which to me seems embarrassing.

The reticence of the Catholic church in this regard puzzles me. The church either has not reacted, or reacted with cautious agreement, to the international advisory texts, the declarations (UNESCO) and conventions (European Council). On the other hand it disciplines theologians and the faithful for reasons that are difficult to defend. Even the suspicion of ambiguity in pregnancy crisis-counselling (obligatory counselling in the three months during which abortion is not penalised) which could (but de facto does not) fall onto church-related counselling services has led to vehement interference with the German bishops by the Pope and the Prefect of the Congregation for the Doctrine of the Faith. Not even a stain of merciful practices is to fall onto the clean, white teaching-robe of the church's interdiction of abortion which is penalised by excommunication. On the other hand, church authorities hardly, or only very indirectly and abstractly, criticise possible mis-developments of biotechnology, that is to say, human genetics. Should the Catholic church, in opposition to the intent of the Pastoral Constitution of Vatican II, move in the direction of developing a split between the message for the world and the message for the church? The inclination only to pay close attention to ethical problems when they are ecclesiastical problems emphasises the march into the Catholic ghetto more than it would happen by a well-argued critique of the 'modern' system of interdependence of research, technology, and economy. However old and wise the church authorities may consider themselves to be, the discussion with 'bioethics' is yet to come; they have not got it over and done with.

2 The Failure of 'Rights Talk'[1] in the Field of Bioethics: The European Convention on Human Rights and Biomedicine

NEVILLE COX[2]

> I know what's legal not what's right and I'll stick to what's legal (Thomas More).[3]

Introduction

The comments of Sir Thomas More, scholar, philosopher and lawyer are indicative of the general tenor of this paper. Other chapters of this book will address issues pertaining to the moral and scientific rights and wrongs of biomedicine. The main question which is addressed here, however, concerns the proper extent to which issues relating to genetics and biomedicine should be regulated by law, and even more especially, the proper moral and legal principles upon which such regulation should be based. The tangible vehicle for this analysis is the *European Convention on Human Rights and Biomedicine*[4] (formerly the *European Bioethics Convention*), which constitutes an attempt to impose some supranational legal control on the practise of biomedicine within Europe.

This paper essentially comprises four sections. First, there is a brief consideration of the Council of Europe, the institution under whose auspices the Convention was drafted, and specifically of the nature of any laws produced by this institution. Secondly, the history and content of the Convention itself will be considered. Thirdly, analysis will be made of the most substantial criticisms which may be made thereof.

The final section, however, is rather more intricate. Focusing on Article

18 of the Convention, which seeks to ensure *adequate protection for the embryo*, in the context of any national law permitting embryonic research, it addresses the question of how such legal protection is to be justified. The Convention on Human Rights and Biomedicine, as its name suggests, answers the question by reference to the language of rights. It is submitted, however, that, like many aspects of the Convention, embryonic protection within the Council of Europe cannot adequately be justified on a human rights analysis. This leads to the conclusion that if matters of controversy within the area of biomedicine are validly to be regulated by law, then such law requires a more wide-ranging language of ethics and morality for its justification. Accordingly, if the Council of Europe is to continue to be involved in these areas then this may necessitate the creation and operation of definite ethical norms in this area.

The Council of Europe

The Council of Europe, the first international organisation to be founded in Europe after the Second World War, was set up in 1949 in the immediate aftermath of that war to ensure greater cooperation between the nations of Europe on a range of issues, social, cultural and, to a lesser extent, political.[5] Although there are connections between the two institutions, it is important to distinguish the Council of Europe from the European Community and European Union, which latter body stems from the old European Economic Community, set up in 1958 to further the economic rebuilding of postwar Europe.[6]

The significance of the distinction, for legal purposes, lies in the fact that European Community law is directly effective in Ireland. In other words, European Community Treaties, Regulations of the Council of Ministers of the European Community[7] and decisions of the European Court of Justice are instantly and directly a part of Irish law, just as if they had been promulgated by an Irish court or the Irish Legislature.[8] Furthermore, European Community Law is superior to Irish law[9] and even the Irish Constitution,[10] so that in the event of a conflict between Irish constitutional law on the one hand and a necessary aspect of European Community law on the other, it is the latter which will win out.[11] The primacy and directly effective nature of European Community law was quite clear in 1973 when Ireland joined what was then the EEC. As a result a constitutional amendment was necessary in order to allow Ireland to enter this new legal order.[12]

In comparison, the law which is generated by the Council of Europe is neither directly effective in Ireland nor superior to Irish law. This is because in respect of all aspects of international law other than European Community law, Ireland adopts what is known as a dualist approach.[13] This means that if an international agreement is to become an actual aspect of Irish law, then three criteria must be fulfilled. Not only must it be signed and ratified by the Irish government, but it must also be incorporated into Irish law by a piece of domestic legislation. This illustrates the fact that all public international law is based on the consent of the parties thereto.[14] Laws of the Council of Europe come within this category and therefore, while influential they are not directly effective in Ireland.[15]

Undoubtedly the most notable contribution of the Council of Europe to a harmonisation of European law[16] has been the *European Convention on Human Rights*[17] *and Fundamental Freedoms.*[18] Providing a centralised understanding and protection of human rights within Europe, this document has paved the way for much social change within contracting states.[19] Thus for example, it is because of the European Convention on Human Rights that Ireland amended its homosexuality laws,[20] its rules relating to civil legal aid[21] and its absolute restriction on information in respect of abortion.[22]

Equally, however, the operation of the European Convention on Human Rights is indicative of a problem which the Council of Europe has in its work generally. There is much diversity of value systems within Europe, and to attempt to find a universally acceptable definition of the precise content of human rights is frankly impossible.[23] The response of the European Court of Human Rights to this difficulty, has been to shy away from situations where it would be required to make controversial decisions.[24] This is achieved by two means. First, the court affords a substantial discretion, known as a *margin of appreciation* to member states in respect of moral matters.[25] Thus, for example it has regularly allowed the right to freedom of expression to be restricted within member states in order to suppress what an individual state regards as blasphemous material.[26] Secondly, it has tended to fudge matters which are politically or morally sensitive. This is particularly notable in the area of abortion, where there is simply no clear European consensus. The court has refused to pronounce definitively as to whether the life of the foetus is included within Article 2 of the Convention (right to life),[27] whether Article 8 (privacy) includes a right to have an abortion,[28] or whether a father objecting to the abortion of what he sees as his offspring has any rights to prevent such abortion.[29]

It is in light of the difficulties in creating uniform moral or ethical norms for states which may have a very different moral ethos that the *European Convention on Human Rights and Biomedicine* must be viewed for it is an attempt to create some degree of legal uniformity in an area which is morally controversial in the extreme. These difficulties are reflected both in the content of the Convention and, more specifically, in the language which it uses to achieve its objectives.

The History of the Convention

From its initial concern with human rights generally, the Council of Europe has become involved in a number of different areas, and has produced international agreements in respect of such areas.[30] The major forces within the Council of Europe, namely the Committee of Ministers[31] and the Parliamentary Assembly,[32] have, since the end of the 1970s considered issues pertaining to the operation of medical science. Since 1976, the Council of Europe has produced recommendations, for example, in respect of the care of the terminally ill,[33] hospital conditions for the mentally ill,[34] developments in biotechnology and consequences for agriculture,[35] the protection of medical data[36] and, on more than one occasion, the rights of individuals in relation to genetic testing and the confidentiality of results of such testing.[37]

Since 1986, the Parliamentary Assembly has focused in particular on ethical issues which arise in respect of medically assisted procreation, and the ways in which the embryo is used in medical research.[38] Recommendations made at this time prohibited any creation of embryos for scientific, commercial or industrial use, any commercialisation of embryos and any research on viable embryos save where this is essential to the survival of the embryo.

Real action in the area of bioethics, however, began in 1985 with the setting up of an ad hoc Committee of Experts on Bioethics (CAHBI), created as an aspect of the legal affairs department and in 1992 given the title of the Steering Committee on Bioethics (CDBI). Composed of 60 experts, including doctors, researchers, jurists and philosophers, its task was to encourage action in the area of bioethics at parliamentary level and, ultimately to produce a European Bioethics Convention.[39]

Work on such a Convention began in June 1991,[40] with a Working Party to prepare to draft the Convention being set up in March 1992. In June 1994 the first draft of the Convention was presented to the Parliamentary Assembly for consideration. It was rejected, however and sent back to the Committee of Experts, largely on the grounds that it did not provide an adequate protection against abuse, particularly in respect of the protection of embryos and the significance of consent in the area of biomedical research.[41]

On 7 June 1996, the Steering Committee on Bioethics, taking account of the above Parliamentary Assembly opinion, produced a new draft, entitled the *Convention on Human Rights and Biomedicine*.[42] This new Convention was adopted by the Parliamentary Assembly on September 26, 1996, albeit with little enthusiasm.[43] There was a general feeling that while it was not very good, it was at least better than nothing, and would serve as a yardstick and model for individual member states.[44] Indeed, as the first international text of its kind, designed to balance competing claims of progress in science and individual human rights, it is arguably reflective of the highest level of consensus possible in this area at this time. On 19 November 1996, it was adopted by the Committee of Ministers.[45] The Convention has been open for signature since 4 April 1997[46] and was signed that day, in Oviedo, Spain[47] by 21 European countries.[48] Ireland has not yet signed it, and according to the Department of Foreign Affairs, there are no immediate plans to do so.[49] For the Convention to come into force there must be five ratifications thereof,[50] of which four must be made by Council of Europe member states.[51] Currently only two states, San Marino[52] and Slovakia,[53] have ratified the Convention. Furthermore, any contracting party may, in signing the Convention, make a reservation in respect of a particular clause.[54] Finally, any party may, at any time, denounce the Convention, by means of notification to the Secretary General of the Council of Europe.[55]

The Significance of the New Title of the Convention

Before analysis is made of the content of the new Convention one final point should be mentioned. As was noted, when the Second Draft of the Convention was put before the Parliamentary Assembly, in June 1996, it was no longer called the European *Bioethics* Convention, and was now known as the Convention on *Human Rights and Biomedicine*. This change is significant for two reasons.

First, by referring to human rights an explicit and intended connection is made between this new Convention and the European Convention on Human Rights,[56] whose jurisprudence is well known to member states.[57] Indeed, under Article 29 of the Convention on Biomedicine, it is provided that the European Court of Human Rights may give advisory opinions on legal questions concerning the interpretation of the Biomedicine Convention (without reference to any specific proceedings pending in a court). Such opinion will be offered at the request either of the government of a member party (after having informed all the other member parties) or at the request of the Steering Committee on Bioethics, newly empowered by Article 32 of the Convention.[58]

The similarities between the two documents are striking, not only in the ideology which supports their underlying principles, but also in terms of the legal notions which are used. For example, the rights protected in both Conventions may be restricted if the same three criteria are respected and followed: that the restriction be prescribed by law,[59] serve a legitimate purpose[60] and be necessary in a democratic society.[61] The most notable similarity between the two documents, however, is ideological. In both cases it is felt that the legal and arguably the moral validity of a course of action may be determined by a human rights based analysis. On this basis, the extent to which the jurisprudence created under the European Convention on Human Rights will be relevant as an aid to understanding the precise ambit of the new Convention remains to be seen.

Secondly, the change in the title of the Convention is significant because by omitting any reference to bioethics explicit change has been made to the mandate of the Steering Committee, from a concern with the uncertain area of ethics to a concern with the far more well known and far safer area of Human Rights. Joseph Raber comments that,[62]

> The disappearance of the concept 'bioethics' from the text is revealing in itself ... For the word encompasses such a wide area and raises such new questions that it may intensify divisions in a Europe where there is already a great mix of cultural philosophies and religious traditions.

This relates to an inherent problem to which reference has already been made – that because different states will have different values, therefore any supranational law dealing with a controversial issue will necessarily only be able to achieve a limited amount. Such laws depend on the consent of those who are to be bound by them, and therefore what will be generated will inevitably be the lowest common denominator of all such systems, in this case the tried and trusted area of human rights.

This leads on to one final point which is the ultimate argument made in this paper, namely that the major flaw within the Biomedicine Convention is that the ethics of science and biomedicine regulated by the Convention should and inevitably will involve more than merely a respect for the rights of others. There are also social values involved which cannot be categorised by the overused language of rights. It will later be suggested that this is particularly important in the context of the protection afforded by the Convention to the human embryo. Before this, however, it is necessary briefly to consider the overall content of the Convention.

The Content of the Convention

Despite the fact that since 1994 the Convention has had a number of false starts and has been the subject of protracted negotiations, and despite the fact that many of its articles have been the subject of modification or amendment, its basic structure has remained the same throughout.[63] It comprises 14 chapters and 38 articles, dealing with the relationship between the needs of research, society and science on the one hand and the rights and dignity of the individual human being on the other. It is clear from Article 2 of the Convention that the interests and welfare of the human being are intended to prevail over the sole interests of society or science, but as shall be noted later, it is at least arguable that parts of the Convention in fact favour claims of science in this context.

The Explanatory Report to the Convention suggests that it has three major and separate areas of concern. First, it is concerned to protect the individual human being from any threat arising from the misuse of science. Secondly, it has a concern for society (and the Explanatory Report seems to imply that this exists both in respect of the social benefits arising from scientific developments and also in respect of the interests of society as a collection of individuals *sharing a number of ethical principles and governed by legal standards*). Finally, it has an interest in *the human species*.[64] According to the Explanatory Report, protection is afforded to the human species throughout the Convention, beginning with the preamble which refers to *future generations* and *all humanity*. It is questionable, however, whether the distinction between the *human species* and the *individual* is anything other than illusory. The preamble also refers to the human being *both as an individual and as a member of the human species*. Moreover, in view of its title, and its overt concern with *human rights and dignity* it is certainly arguable that the Convention's principal ethical concern is with the interests of individual humans, and the significance of the *human species* derives exclusively from the fact that it is the receptacle of these individual members.

The Preamble to the Convention

The Convention's concern with human rights, is set out from the outset in its preamble. Reference is made to the Universal Declaration of Human Rights (1948), the European Convention on Human Rights (1950), the European Social Charter (1961), the International Covenants on Civil and Political and Social and Cultural Rights (1966) the Convention for the Protection of Individuals with regard to Automatic Processing of Personal Data (1981) and the Convention on the Rights of the Child (1989).[65] It notes the signatories' concern to protect the rights of the human being, both as an individual and as a member of the human species, and also to protect the dignity of the individual, which may be deleteriously affected by misuses of biology or technology. On the other hand, it also notes the interest of the signatories in the promotion of the sciences of biology and medicine and the benefits deriving to society from progress in these sciences. The objective of the Convention is to create a reasonable reconciliation between these two concerns, which reconciliation is only possible, ... *if public awareness is aroused as a result of an international instrument devised by the Council of Europe in line with its vocation.*[66]

Medical Interventions

As a general rule,[67] the Convention provides that where a person is capable of giving consent,[68] he or she must do so in advance of any medical intervention.[69] Moreover, the patient must be fully informed[70] of the purpose and nature of the medical intervention as well as of its consequences and risks.[71] Finally, he or she is entitled to withdraw his or her consent at any time.[72] The only situation in which this does not arise is where, by reason of emergency, a person requiring medical intervention cannot consent to such intervention, for example where he or she is in a coma. In such cases, intervention may be carried out immediately for the benefit of the health of the individual concerned.[73]

The above obviously raises the question of what happens in the case of a person who, because of legal incapacity, mental illness, or general medical state is unable to give or to withhold consent.[74] In the case of a minor, medical treatment may only be given for the direct benefit of that minor and then only where permission has been given either by his or her representative or by an authority provided for in law. Moreover, the opinion of the minor is taken into consideration and becomes increasingly important with his or her increasing age and maturity.[75] Similar considerations apply in respect of an adult unable to consent to a medical intervention,[76] with such an adult taking part in the authorisation procedure *as far as possible*.[77] Finally,[78] in the case of a person who has a mental disorder of a serious nature, an intervention is permitted where it is aimed at treating his or her disorder[79] and where, in the absence of such treatment, serious harm to his or her health is likely to arise.[80]

Medical and Scientific Research

The same approach is taken in respect of medical and scientific research involving humans.[81] Under the Convention, scientific research in the field of biology and of medicine shall be carried out freely, subject to the provisions of the Convention and other legal provisions ensuring the protection of *the human being*.[82] In order to justify the use of scientific research on a person, five criteria must be fulfilled:[83]

- there must be no possible alternative procedure to research on humans which is of comparable effectiveness;
- the research project must have been approved by *the competent body* after a review both of its scientific merit *and* of its ethical acceptability;
- the risks incurred by the human subject of the research must be proportionate to the benefits accruing to society from the research;
- the individual must be informed of all his or her legal rights in respect of this issue and of any legal safeguards which exist in that area; and
- most importantly, having been so informed, the individual must consent to the relevant procedure. Such consent must be express, specific[84] and documented.[85]

The Convention goes on to deal with situations where the subject is incapable of consenting to research.[86] It is stressed that in such circumstances research shall only be permissible where a number of criteria are fulfilled. First, the research in question must have the potential to produce real and direct benefits to the health *of the subject*. Secondly, the subject of the research must not object to it.[87] Thirdly, research of comparative effectiveness must not be able to be carried out on persons capable of consenting to such research.[88] Finally, all the relevant conditions listed elsewhere in the Convention must be fulfilled.[89]

More controversially, paragraph two of Article 17 permits research '[e]xceptionally and under the protective conditions prescribed by law', even where the research will not produce direct benefits to the health of the person concerned. In order for this to be permitted, all the relevant conditions in Article 17 paragraph one must be fulfilled, and furthermore it must be the case that:[90]

- the research has the aim of contributing, through significant improvement in the scientific understanding of the individual's condition, disease or disorder, to the ultimate attainment of results capable of conferring benefit to the persons concerned or to other persons in the same age category or afflicted with the same disease or disorder or having the same condition; and
- the research entails only minimal risk and minimal burden for the individual concerned.[91]

The Explanatory Report to the Convention argued that were such research to be absolutely banned,

> ... progress in the battles to maintain and improve health and to combat diseases only afflicting children, mentally disabled persons or persons suffering from senile dementia, would become impossible.[92]

Interestingly, it would seem that despite its professed concern with *individuals* the Convention is prepared to categorise such individuals into various groups – minors, mentally disabled persons, etc. – and to imply that as a member of such a group the individual has an interest in the well-being of all other members of that group. It is on this basis that the Explanatory Report regarded such research as possibly acceptable.

Organ and Tissue Removal and Transplantation[93]

The Convention also looks at the issue of organ and tissue removal for transplantation purposes.[94] Under Article 19, such procedure may only be performed on a living person, where that person has consented under the general rule in Article 5 of the Convention,[95] where there is no suitable organ or tissue available from a deceased person, where there is no alternative therapeutic measure of comparable effectiveness[96] and where the transplantation is carried out solely for the therapeutic benefit of the recipient.

Article 20 deals with persons who are unable to consent to organ removal. In principle the procedure of organ removal in respect of such persons is prohibited.[97] On the other hand, in exceptional circumstances, under the protective conditions laid down by law, and where what is being transplanted is *regenerative tissue*,[98] the removal of such tissue from a person unable to consent to such removal may be authorised. Once again there are stringent conditions laid down which must be complied with if such procedure is to be permitted, namely:

- there is no compatible donor available who has the capacity to consent;
- the potential donor does not object to the procedure;[99]
- the authorisation under Article 6, to which reference has already been made, must be given in writing;
- the donation must have the potential to be life saving for the recipient and hence, by implication, the situation must involve a threat to his or her life; and
- the recipient and donor must be siblings.[100]

This section attracted a good deal of criticism from commentators who felt that it gave precedence to the claims of science over those of the individual. Specifically there is criticism of the fact that persons who are unable to consent may be used as organ donors *inter alia* if there is no compatible donor available who has the capacity to consent. The Convention is seen as a victory for science over the constraints of ethics. It may, however, be argued that this is a rather melodramatic response and that the constraints in place will mean that such 'non-consensual transplants' will occur only rarely and in limited factual circumstances.

Miscellaneous Sections

The Convention also requires the provision of equitable access to health care,[101] and adherence to professional standards in the health field.[102] It outlaws the use of the human body and its parts as a source of financial gain.[103] It insists that, as a matter of right, individuals are entitled to respect for private life in relation to information about their health[104] and it recognises a right to information about one's health,[105] which right may, in exceptional circumstances, be restricted.[106]

In its consideration of the human genome,[107] the Convention bans any interference with or modification of the genetic line of a subject save where such intervention occurs for preventative, diagnostic or therapeutic purposes.[108] It also forbids any use of techniques of medically assisted procreation[109] for the purpose of choosing a future child's sex, except where there are serious hereditary sex-related diseases to be avoided.[110] Furthermore, any form of discrimination against a person on the basis of his or her genetic heritage is prohibited.[111] Finally, the Convention bans all tests predictive of genetically related diseases, save where they are being carried out for health purposes or for scientific research linked to health purposes.[112]

Chapter VIII of the Convention, which deals with infringements thereof, has given rise to a good deal of complaint in that it is argued that there is no adequate mechanism provided by the Convention, through which its provisions may be enforced. It is up to the laws, and specifically the judicial branch of government, of contracting parties to determine how the Convention is to be enforced, by 'prevent[ing] or ... put[ing] a stop to an unlawful infringement of the rights and principles set forth in this Convention'.[113] Signatory states must also provide for appropriate sanctions to remedy the breach of these provisions[114] and for compensation in the event of such breach.[115]

The European Court of Human Rights is mentioned, but only for the purposes of giving an advisory opinion in respect of the interpretation of the Convention, not for providing a statement that a contracting party has breached the Convention.[116] Moreover, such an opinion may only be sought by a member state, or by the Steering Committee on Bioethics. This contrasts sharply with the procedure under the European Convention on Human Rights where any individual 'victim' of Human Rights abuse may appeal to the Court and Commission of Human Rights for a statement that his or her rights have been breached and for appropriate reparation in respect of such breach.[117] The absence of an effective enforcement procedure is, therefore, a valid criticism of this Convention, but equally it should be remembered that, being based on international law, it will necessarily be more dependent on consent than on enforcement.

The Additional Protocol to the Convention on the Prohibition of Cloning Human Beings

The Convention envisaged the development of the principles contained in the Convention in specific areas by means of Additional Protocols to the parent convention.[118] Originally, it was envisaged that protocols would be enacted in the areas of organ transplants,[119] genetics,[120] medical research and the protection of the human embryo and foetus. On 5 July 1996, however, these plans were overtaken by the 'birth' at the Roslin Institute in Scotland of Dolly, a sheep cloned from the cell of an adult ewe. News of this cloning became public in February 1997[121] and whereas there was a good deal of scientific support for the breakthrough, equally it was seen by some as being deeply morally repugnant. Far worse, however, for many was the prospect which the whole *Dolly* episode seemed to foreshadow – the possibility of human cloning.

On 25 February 1997, the Secretary General of the Council of Europe, Mr Daniel Tarschys announced his extreme opposition to the possibility of human cloning.[122] He insisted that 'The cloning of an adult sheep may be an impressive scientific achievement but it also demonstrates the need for firmer rules on bioethics ... No human cloning whatsoever is acceptable'.

Mr Tarschys further announced that work would shortly begin on a Protocol on the Protection of the Embryo and Human Foetus which would include a ban on human cloning.[123] In fact, when the Protocol was produced for signature in Paris on 12 January 1998,[124] it focused exclusively on the Prohibition of Cloning of Human beings.[125] On that day, 19 states signed the Protocol, which is the first binding international treaty on the subject of cloning human beings.[126] Speaking at the signing ceremony, Secretary General Tarschys announced that

> At a time when occasional voices are being raised to assert the acceptability of human cloning and even to put it more rapidly into practice, it is important for Europe solemnly to declare its determination to defend human dignity against the abuse of scientific techniques.[127]

The issue of human cloning raises questions both legal and ethical which are beyond the scope of this paper.[128] For present purposes it is proposed briefly to note the content of the new protocol.

The Additional Protocol was deemed to build on certain provisions of the Convention, specifically: Article 1, which refers *inter alia* to the *dignity*, *identity* and *integrity* of all human beings; Article 13 which prevents research based modifications of the human genome; Article 18 which protects the embryo in the context of research and which prevents the creation of embryos for research purposes.[129] The preamble to the Additional Protocol recognises the potential benefits to science of cloning techniques, but also sees the possibility of human cloning as posing a threat both to individuals involved and to human dignity and hence amounting to a misuse of biology and medicine. On this basis the Protocol, which comprises eight articles, seven of which are essentially procedural, asserts that *any intervention seeking to create a human being genetically identical*[130] *to another human being whether living or dead is prohibited.*[131]

In fact the framers of the protocol envisaged three different situations in which cloning might occur; the first, the cloning of cells as a technique was deemed to be entirely ethically acceptable.[132] The second, the use of embryonic cells in cloning techniques was deemed to come under the auspices of the forthcoming protocol on embryonic protection. The final situation, the cloning of human beings by either embryo splitting or nuclear transfer, was the focus of this new Convention.

The justification for the prohibition on human cloning as set out in the Explanatory Report to the Additional Protocol is interesting. First, cloning is seen as a threat to *human identity*, because the human genetic condition might be predetermined by a third party. Secondly, human dignity is seen as being *endangered by instrumentalisation through artificial human cloning*. Finally, it was felt that the ban on human cloning could be justified by reference to the individual human being on the basis that '... naturally occurring genetic recombination is likely to create more freedom for the human being than a predetermined genetic make up'. It will later be argued that none of these justifications adequately explain the existence of the protocol just as the concept of human rights does not explain the existence of the parent Convention.

One significant point remains. For the purposes both of the Convention on Biomedicine and of the Additional Protocol, the term 'human being' is left open for definition by domestic laws of signatory parties.[133] Clearly this is of huge significance in respect of the embryo and the foetus. It may be argued that if a particular member state does not regard a 'human being' as existing until the moment of birth, then that state might validly permit human cloning by the technique of embryonic splitting, while not violating the terms of the Convention. The precise impact of this point remains to be seen.

Embryonic Research and the Convention on Biomedicine

Despite the increasing global concern with embryonic research,[134] the Convention's reference to such research consists of one article divided into two subsections.[135] These provide first, that where the law allows research on embryos *in vitro*, it shall ensure adequate protection of the embryo[136] and secondly that the creation of human embryos for research purposes is prohibited.[137] This reference to what is clearly an important issue may seem rather brief, but it was proposed that the debate on embryos generally would be more fully dealt with in an Additional Protocol on the Protection of the Embryo and Foetus. Given, however, that arguments about embryonic research were a major stumbling block to be overcome in the creation and acceptance of the original Convention,[138] such a protocol will, presumably, be difficult to formulate and unlikely to gain significant widespread support.[139]

The fundamental concern of this paper, however, is not the extent to which the Convention seeks to protect embryos, but rather the fact that it does so at all, or, more precisely, the fact that it justifies such protection by reference to very specific terminology. In its preamble and in its opening articles the Convention stresses that it exists as a protector of the dignity of the *human person* and of the *rights of persons*. Moreover, Article 18 occurs in Chapter V of the Convention (Scientific Research). Article 15 of the Convention sets out a general rule for this Chapter, namely that scientific research shall be carried out freely subject to legal provisions *ensuring the protection of the Human Being*. In similar vein, the Committee of Ministers in their resolution[140] authorising the CAHBI to draw up a Convention in this area referred to human dignity once and human *rights* three times. The clear implication is that the principled justification for Council of Europe regulation of embryonic research is human rights, or at least human person based. Given that embryonic research, however, operates on the basis of the consent of donating parties the human right or human dignity operative in this context must belong to the individual embryo upon whom research is being performed.

This is a startling prospect if for no other reason than, as has been mentioned, in its interpretation of the European Convention on Human Rights in respect of abortion the Court of Human Rights has never suggested that the unborn, in any category, might be classifiable as 'human beings' or enjoy a right to life.[141] This is based both on a desire not to get involved in matters which are politically or morally sensitive and also a feeling that it would be wrong for a supranational court to impose what the American Supreme Court in the case *Roe v. Wade*[142] referred to as 'one theory of life' across a Europe where the significance attaching to the unborn varies from country to country[143] and where judicial acceptance of such a 'theory of life' would effectively foreclose argument in what is a most controversial political issue.[144] For the Convention on Biomedicine to afford the *in vitro* embryo a supranational right to life or to imbue it with personhood is, therefore, problematic for two reasons, both of which shall now be considered. First, it is an uncharacteristically drastic answer to a question which, in global terms, is entirely unresolved and probably unresolvable, namely the question whether humans have legal rights before birth. Secondly, it displays an inconsistency within the Council of Europe's approach to unborn life which is reflective of the fact that if it wishes to deal with matters pertaining to bioethics, including embryonic research and human cloning, the Council will have to be prepared to depart from the tried and trusted, yet now inadequate language of rights and use the far more difficult language of morality.

The Drastic Step of Affording the Unborn Legal Rights

It is vital from the outset to distinguish the philosophical/theological/medical question of when life begins from the legal question of when the law begins to recognise and protect such life. Although the lawyer may be guided by scientific, medical and theological factors, when he or she asks the question 'when does life begin' he or she is essentially asking, if and when, on the basis of reason and policy, the law will start to afford legal protection or, even more dangerously legal rights, to the individual unborn entity.

This is a question which simply cannot be resolved by recourse to easy logical propositions.[145] Glanville Williams, the celebrated English jurist correctly observed that there is no real consensus, medical or philosophical, as to when life begins and hence there is no point along the line from fertilisation to birth where introduction of legal protection for the unborn would not seem arbitrary.[146] This does not mean that protection of unborn life should not be introduced at all, but it does lead to the view that after all relevant moral considerations have been taken on board, the ultimate question for any law in this area should, of necessity be policy based, concentrating on the practicalities of affording legal protection to an individual unborn life at any particular stage before birth. Thus for example, if the constitutional protection afforded under Irish law[147] to the 'unborn' was deemed to operate from the moment of fertilisation, then this would clearly render use of contragestive or abortifacient forms of birth control, such as the morning after pill or inter-uterine device unconstitutional.[148] Accordingly, it might be good policy for Irish law to protect the 'unborn' no earlier than the point of implantation.[149] In similar vein, the US Supreme Court in the seminal decision *Roe v. Wade*,[150] argued that 'unborn life' would only become a compelling interest[151] capable of being balanced against the mother's fundamental right to abortion at the point of foetal viability, or more precisely at the beginning of the third trimester of pregnancy.[152] Commentators have noted the impact of policy here, such that at the point of viability within the American scheme of things, all relevant rights and interests would appear to be balanced, and hence at this time policy dictates that the foetus may enter forcibly into the legal equation.[153]

The German law in this area is instructive.[154] Largely because of its history in which the value of human life was, for a period, determined by the nature of that human life, the German Basic Law or Constitution recognises a right to life of the unborn from the fourteenth day after fertilisation.[155] On the other hand it treats abortion by social mechanisms (counselling) rather more than with the criminal law.[156] Equally, it is particularly keen to restrict *in vitro* embryonic research.[157] Its 1991 Law on the Protection of Embryos defines the embryo as inviolable from the beginning of its existence (without affording that entity a right to life). The production of embryos exclusively for research is prohibited. The only goal for the fertilisation of the ovum can be the pregnancy of the woman from whom the egg came. The law prohibits the use of the embryo to any other end before implantation, and any research which involves more than simple observation is prohibited.

The reason why this is significant for present purposes is as follows. If, under the Convention on Biomedicine an embryo is, as a matter of morality and policy, to be protected on the basis that it is recognised as a person or a possessor of rights, then presumably protection would need to be along the lines of the German law. Yet this is not the case. In fact, the Convention actually presumes the legitimacy of laws which permit *in vitro* embryonic research. The vague reference to the protection of the embryo is merely something that must happen when such research is ongoing. If the embryo really *is*, as a matter of policy and law a possessor of human rights under this Convention, then the scant protection which it receives in Article 18 is a manifest violation of Article 2 of the Convention which, after all, urges that the 'human being' is more important than the sole interests of science.

This indicates either a major inconsistency between Article 2 and Article 18 of the Convention, or more probably, given the traditional Council of Europe reluctance to pronounce on the legal nature of pre-birth life, that the Steering Committee on Bioethics did *not* see each individual embryo as being of sufficient status that it could legitimately be afforded rights, yet did feel that wholesale destruction of embryonic life for the cause of science was, somehow, deserving of legal censure, not because it violated any human rights, but because it was ethically suspect.[158] Similarly, it may be argued that what is at threat in the area of human cloning is *not* the rights or dignity of any human being but is rather *the ethics of biomedicine*. The problem for the framers both of the Convention and of the Additional Protocol was that, in attempting to find common ground for this supranational treaty, they were forced to abandon the untried language of ethics in favour of the tried and trusted, yet inappropriate language of rights.

The reason why such language is inappropriate is not because of the moral status of the embryo, but because of the legal nature of rights.[159] Despite their overuse in this century, rights are, at a fundamental level, very specific things. Their recognition, albeit not their application, should transcend all reasonable moral controversy. The alternative involves the use of 'rights talk' to conclude and foreclose debates in difficult areas, and as a concomitant, it means that 'real' rights – of freedom of expression, property and so on – are debased. In proper form, the recognition of a right to do something involves an implicit statement that absent such right the state involved is not fulfilling the requirements of a democracy.[160] To afford such a gigantic legal status to the embryo, an entity whose moral status, far from being above moral controversy actually generates it, is certainly dangerous and arguably wrong.

On the other hand, it is entirely legitimate to be edgy about the concept of mass destruction of foetuses, of cloning, of interference with genetic lines and of destructive embryonic research, not because rights are being violated, but because of the ethical implications, both religious and secular of scientists attempting to 'play God'. Indeed it is arguable that such behaviour should be controlled, or even prohibited by law, simply because of its ethical unacceptability.[161] The difficulty for the Council of Europe and other states and institutions who wish to invoke such control is the principled basis on which such control is to depend.

The question of whether it is legitimate to enforce by law certain perceived moral or ethical norms *because of the content of those norms* and without reference to the effect of their violation on other people has been debated from the time of Aristotle to the present day. Such enforcement has been justified by reference to the duty of a state to make its citizens moral,[162] the duty of a state to ensure that its citizens earn eternal salvation,[163] the obligation on citizens to obey society's rules even when they aren't harming someone[164] and the importance to social cohesion generally of a shared idea of the difference between right and wrong in that society.[165] While such arguments are, perhaps unfashionable when compared to the view of John Stuart Mill[166] that the only reason for using the law to require a person to behave in a particular way was to prevent *harm to others*,[167] this is not a reason why they are invalid. More particularly, as Lord Patrick Devlin has demonstrated, a great number of important laws, can be justified only as instances of the *enforcement of morals*.[168] Thus, if the Council of Europe wished to regulate the ethics of biomedicine by law, there is no shortage of traditional principled justifications for doing so.

The problem for Lord Devlin and others like him, however, was not that he could not justify the enforcement of a society's morality but rather that he found it difficult to give a workable test for determining what that morality was[169] and, bearing in mind the significance of individual liberty, for adducing those aspects of that morality which were so important that they merited legal protection.[170] It has been argued that Lord Devlin's test would involve, not only the enforcement of morality, but also the enforcement of prejudice.[171]

This is also the major problem for the Council of Europe in its dealings in this area. It has no certain basis apart from the language of rights upon which to make moral judgments. What is needed, as the Council of Europe moves in new directions such as the direction of bioethics is some effort to determine a moral or ethical code[172] for Europe in this area. This task may, particularly in light of European religious heterogeneity, be unachievable. Nonetheless, if the Council of Europe is to exert any worthwhile control over the area of Biomedicine, it is a step which will shortly have to be taken. The Council of Europe must expand its moral web beyond the confines of human rights and learn to speak the increasingly important language of ethics.

Notes

1 For a general criticism of the overuse of the concept of rights as a means of resolving all political controversies see Glendon, 1993.

2 My thanks are due to a number of people who have assisted me in respect of this piece. Particularly I would like to thank Professor William Binchy and Mr Eoin O'Dell of Trinity College Dublin. I am also grateful to Mr Allen Figgis and Ms Sharon Smith.

3 See Bolt, 1960, p. 73.

4 Hereafter 'The Convention'.

5 The Council of Europe grew out of the Hague Congress of 7 May 1948, better known as the Congress of Europe. The Congress was attended by more than a thousand delegates from over 20 countries. It produced resolutions calling *inter alia* for the creation of an economic and political union to guarantee security, economic independence and social progress; the establishment of a consultative assembly elected by national parliaments; the drafting of a European Charter of Human Rights and the setting up of a Court to enforce its decisions. On 5 May 1949 in St James's Palace, London, the Treaty constituting the Statute of the Council of Europe was signed by Belgium, France, Luxembourg, The Netherlands, the UK, Ireland, Italy, Denmark, Norway and Sweden. Under Chapter 1 of this Statute the objective of the New Council was deemed to be '... to achieve a greater unity between its members for the purpose of safeguarding and realising the ideals and principles which are there common heritage and facilitating their economic and social progress'. Between 1949 and 1970, eight more states joined the Council of Europe and by February 1996, when the Russian Federation joined the Council, membership stood at 40 states encompassing more than 700 million citizens.

6 The distinction is difficult, in that both entities operate on the basis of a court (the European Court of Human Rights as against the European Court of Justice), both have a parliament, and both use what is essentially a Committee of Ministers. Moreover the European Union is governed by a Council, and to further complicate matters, all the member states of the EU are also members of the Council of Europe. Finally, the EC has long claimed a human rights jurisprudence. See *Internationale Handelsgesellschaft GmbH v. Einfuhr-und-Vorratsstelle fur Getreide und Futtermittel* (Case 11/70) [1970] ECR 1125, and also *SPUC v. Grogan* (Case C-159/90) [1991] ECR I-4685; [1991] 3 CMLR 849.

7 EC Directives are also directly binding on Ireland in respect of the result to be achieved, but in this case the Irish government has a discretion as to the means by which such result will actually be achieved. See Mathijsen, 1995, pp. 195ff. In the Irish context see *Meaghar v. Minister for Agriculture and Food* [1994] 1 IR 329.

8 *Van Gend en Loos v. Nederlandse Belastingenadministratie* [1963] ECR 1. See also *Murphy v. An Bord Telecom* [1989] ILRM 53.

9 *Van Gend en Loos* cited *supra.* See also *Costa v. ENEL* [1964] ECR 585.

10 *Internationale Handelsgesellschaft* cited *supra.*

11 *Lawlor v. Minister for Agriculture* [1990] 1 IR 356, *Greene v. Minister for Agriculture* [1990] 2 IR 17 and *Meaghar v. Minister for Agriculture* [1994] 1 IR 329

12 Article 29.4.5 of the Irish Constitution. Following the case *Crotty v. An Taoiseach* [1987] IR 713, it would appear that any EC measure which involves a radical departure from the original objectives of the EC Treaty (Treaty of Rome) 1958, must, in order to become part of Irish law, be endorsed by the Irish people through constitutional amendment.

13 On this point see Byrne and McCutcheon, 1996, pp. 678–9.

14 Thus under Article 29.3 of the Irish constitution, it is provided that 'Ireland accepts the generally recognised principles of international law as its rule of conduct in its relations with other states'.

15 For a suggestion that the European Convention on Human Rights can be used as an assistance in the interpretation of the Irish Constitution see *Desmond v. Glackin* (No. 1) [1993] 3 IR 1 *per* O'Hanlon, J.

16 The Council of Europe has also created a European Social Charter (1961), a European Convention for the Prevention of Torture (1987), a European Cultural Convention (1954) and a European Data Protection Convention. Of these, only the Data Protection Convention has been incorporated into Irish law. The Council of Europe also has a Social Development Fund and has impacted in the areas of media and communication, education, sport, and the environment.

17 As shall be noted later, the naming of the document under consideration as the European Convention on *Human Rights* (emphasis added) and Biomedicine has been seen as a reference to this more famous international agreement.

18 This was signed in Rome on 4 November 1950 and came into force on 3 September 1953.

19 The document is unusual, in that not only can an aggrieved individual take an action against his or her member state in respect of human rights violations, but one member state can also take an action against another member state, if it sees abuses going on in that state. See *Ireland v. UK* (1978) 2 EHRR 25.

20 *Norris v. Ireland* (1988) 13 EHRR 186.

21 *Airey v. Ireland* (1979) 2 EHRR 305.

22 *Open Door Counselling and Well Woman Clinic v. Ireland* (1992) 14 EHRR 131 (1993) 15 EHRR 244.

23 A stark example of this concerns the death penalty, and the question of whether the right to life, recognised within Article 2 of the Convention should include a right to freedom from what is, in many countries, common practice.

24 See Hogan (1994), p. 105 for the view that

> To date, the approach of the [European Commission on Human Rights] in cases presenting it with abortion issues has been largely hesitant. The commission, conscious perhaps of the wide divergence of views within the contracting states has seemed anxious to leave as many issues open as possible.

25 See for example *Norris v. Ireland* (1991) 13 EHRR 186, *Dudgeon v. UK* (1982) 4 EHRR 149, and *Handyside v. UK* (1979) 1 EHRR 737.

26 *Lemon v. UK* (1983) 5 EHRR 123, *Otto Preminger Institute v. Austria* (1995) 19 EHRR 35 and *Wingrove v. UK* (1997) 24 EHRR 1. See also Cox, 1997, pp. 101ff.

27 *Paton v. UK* (1981) 3 EHRR 408. The significance, for present purposes of the reluctance of this Council of Europe institution to pronounce definitively on the question of when, for its purposes, life is to be deemed to begin, will be discussed later.

28 *Bruggemann and Scheuten v. Germany* (1981) 3 EHRR 244.

29 *Paton v. UK*, cited *supra.*

30 See earlier at note 16.

31 Resolution (78) 29 on the Harmonisation of Legislation of Member States relating to Removal, Grafting and Transplantation of Human Substances.

32 Recommendation 1159 (1991) (Parliamentary Assembly) on the Harmonisation of Autopsy rules.

33 Resolution 613 (1976) and Recommendation 779 (1976) (Parliamentary Assembly) on the Rights of the Sick and the Dying.

34 Recommendation R (83) 2 (Committee of Ministers) concerning the Legal Protection of Persons suffering from Mental Disorder placed as Involuntary Patients and Recommendation 818 (1977) (Parliamentary Assembly) on the Situation of the Mentally Ill.

35 Parliamentary Assembly Recommendation 1213 (1993).

36 Committee of Ministers Recommendation R (97) 5.

37 Recommendation R (90) 3, on Medical research on Human Beings, Recommendation R (90) 13, on Prenatal Genetic Screening, Diagnosis and Associated Genetic Counselling (Committee of Ministers) and Recommendation 934 (1982) on Genetic Engineering (Parliamentary Assembly).

38 Recommendation 1046 (1986) (Parliamentary Assembly) on the Use of Human Embryos and Foetuses for Diagnostic, Therapeutic, Scientific, Industrial and Commercial Purposes and Recommendation 1100 (1989) (Parliamentary Assembly) on the Use of Human Embryos and Foetuses in Scientific Research.

39 See for example the 1988 CAHBI Report on Human Artificial Procreation.

40 See also the Resolution no. 3 on Bioethics produced at the 17th Conference of European Ministers of Justice, (Istanbul, 5–7 June 1990). At this conference, following the proposal of Ms Catherine Lalumière, then Secretary General of the Council of Europe, Resolution no. 3 on Bioethics was adopted which instructed the CAHBI to examine the possibility of preparing a framework Convention 'setting out common general standards for the protection of the human person in the context of the development of the biomedical sciences'. See also the Parliamentary Assembly Recommendation 1160 (1991) on the Preparation of a Convention for Bioethics. The Assembly recommended that the Committee of Ministers 'envisage a framework Convention comprising a main text with general principles and additional protocols on specific aspects'. In September 1991, the Committee of Ministers instructed the CAHBI

> to prepare, in close co-operation with the Steering Committee for Human Rights (CDDH) and the European Health Committee (CDSP) … a framework Convention open to non-member states, setting out common general standards for the protection of the human person in the context of the biomedical sciences and Protocols to this Convention, relating to, in a preliminary phase: organ transplants and the use of substances of human origin; medical research on human beings.

See the Explanatory Report to the Convention on Human Rights and Biomedicine [hereafter the Explanatory Report], *http://wwww.coe.fr/oviedo/rapporte.htm*, p. 2.

41 Opinion No. 184 of 2 February 1995, Doc. 7210.

42 Denis Durand de Bousingen (1996) comments that,

> Many Parliamentarians have argued from the start that parts of the text needed strengthening and the report given by Giano-Reto Plattner (Switzerland) as well as the amendments proposed by other members, like Austrian MP Walter Schwimmer, helped to ensure, when the big moment came, that the decisive vote went strongly – though still not unanimously – in its favour. As many speakers pointed out in the debate, the Convention was urgently needed, since the total absence of European Legislation on biomedicine left a dangerous legal vacuum in this area. Of course, many states already had stringent regulations, but others were merely feeling their way.

43 Opinion of 26 September 1996, Doc. 7622. Germany, Belgium and Poland abstained when the Committee of Ministers took the vote on the adoption of the Convention. Germany, Belgium and Ireland abstained in respect of the vote on the authorisation of the publication of the Explanatory Report.

44 Thus former Assembly President Miguel Angel Martinez (Spain) pointed out that

> People thought that the European Convention on Human Rights wasn't perfect either, when it first came in. Now its a basic reference tool, not just in Europe, but all over the world.

45 Council of Europe Secretary General Daniel Tarschys said of the newly adopted Convention, 'The text is destined to become the common standard in Europe in the field of Bioethics'.

46 On that day, Secretary General Tarschys said,

> The Convention being opened to signature today is an important achievement. Today, research is a transnational activity and Western companies can carry out research in any part of the world. That is why it is not enough for developed countries to enact laws on their own territory. It needs the broadest agreement possible, and the measures set out in this Convention can exert an influence beyond Europe's boundaries. With this Convention and the protocols which will follow, the Council of Europe has become a privileged forum for debate and action in bioethics matters, and has made itself an active partner in all action in the bioethics field throughout the world.

47 The opening ceremony was attended by the President of Spain, Sergio Marques Fernandez, the Spanish Health Minister, José Manuel Romay Beccaria, representatives of member states and also by the highest dignitaries of the council of Europe.

48 The following countries signed the Convention on 4 April 1997: Denmark, Estonia, Finland, France, Greece, Iceland, Italy, Latvia, Lithuania, Luxembourg, The Netherlands, Norway, Portugal, Romania, San Marino, Slovakia, Spain, Slovenia, Sweden, Turkey, the former Yugoslav Republic of Macedonia. On 6 April 1997, the Convention was also signed by Moldova. Most recently, on 24 June 1998, the Convention was signed by the Czech Republic.

49 At the time when the Convention was signed it was envisaged that the principles contained therein would be developed in additional protocols. See Chapter XII, Article 31 of the Convention. The only such protocol created to date, and one which will be examined shortly is the Additional Protocol in respect of the Prohibition of Human Cloning. However,

additional protocols are being drafted to deal with medical research, organ transplants, the protection of the foetus and embryo and genetics.

50 See Article 33 (3) of the Convention.

51 The Convention is open for signature both to Council of Europe member states and also to nonmember states which took part in its drafting (the Holy See, the USA, Canada and Japan) and finally to the European Community.

52 San Marino ratified the Convention on 20 March 1998.

53 Slovakia ratified the Convention on 15 January 1998.

54 Article 36 of the Convention.

55 Article 37 ibid.

56 The European Convention on Human Rights is referred to in the preamble to the Convention on Biomedicine as also are the *Universal Declaration of Human Rights,* the *International Covenant on Civil and Political Rights,* the *International Convention on Economic and Social Rights,* the *Convention on the Rights of the Child* and the *European Social Charter.*

57 In the authorised Explanatory Report to this Convention, it is noted that

> The term 'human rights' refers to the principles laid down in the Convention for the Protection of Human Rights and Fundamental Freedoms of 4 November 1950, which guarantee protection of such rights. The two Conventions share not only the same underlying approach but also many ethical principles and legal concepts. Indeed this Convention elaborates some of the principles enshrined in the European Convention on Human Rights and Fundamental Freedoms.

58 It is notable that individuals who feel that their rights under the Biomedicine Convention are being violated have no right of petition to the European Court of Human Rights. However, according to the Explanatory Report,

> ... facts which are an infringement of the rights contained in this Convention, may be considered in proceedings under the European Convention on Human Rights if they also constitute a violation of one of the rights contained in the latter Convention.

59 It has been a matter of some controversy as to whether a decision by the highest court in a particular state constitutes 'law' for the purposes of validating a restriction on a listed right. *Sunday Times v. UK* (1980) 2 EHRR 245, but also the European Commission on Human Rights decision in *Open Door Counselling v. Ireland* (1992) 14 EHRR 131. It would, however, seem that the prevalent view is that such decisions do suffice to make the restriction 'prescribed by law'. See the European *Court* of Human Rights decision in *Open Door Counselling v. Ireland* (1993) 15 EHRR 244.

60 These legitimate purposes comprise the preservation of public safety, the prevention of crime, the protection of public health, or the protection of the rights or freedoms of others.

61 Essentially this is a narrow proportionality test, requiring a balance between the restriction and the aim sought to be achieved. See the Court of Human Rights decision in *Open Door Counselling* (above). See also Jowell and Lester, 1988, pp. 52ff.

62 See Raber, 1997.

63 See de Bousingen, 1996, p. 2.

64 Thus the Explanatory Report argues that 'It is no longer the individual or society that may be at risk [from misuses of scientific developments] but the human species itself'.

65 Thus the Explanatory Report comments that

> The principles enshrined in these instruments remain the basis of our conception of

human rights; hence they are set out at the beginning of the preamble to the Convention, of which they are the cornerstone.

66 Ibid. at p. 5.
67 Article 5 of the Convention.
68 The Explanatory Report states that such consent may be either express or implied depending on the nature of the treatment. Thus, whereas generally implied (albeit informed) consent is acceptable, in some cases, such as invasive diagnostic acts or treatment, express consent may be required. Furthermore, under Articles 16 and 19, express consent is required in respect of research or of the removal of body parts for transplantation purposes.
69 According to the Explanatory Report,

> The word, 'intervention' is understood in its widest sense ... – that is to say, it covers all medical acts, in particular interventions performed for the purpose of preventative care, diagnosis, treatment, rehabilitation, or research.

70 It is envisaged that this information must be suitably worded for the person who is to undergo the medical intervention. In other words, he or she must understand the terms used, and must be able on their basis to weigh up the benefits and the detriments of the proposed intervention. See ibid.
71 See ibid. for the view that

> Information on the risks involved in the intervention or in alternative courses of action must cover not only the risks inherent in the type of intervention contemplated, but also any risks related to the individual characteristics of each patient, such as the age or the existence of other pathologies. Requests for additional information made by patients must be adequately answered.

72 The Explanatory Report does, however, state that the withdrawal of a patient's consent *during* an operation need not always be followed, for example where, in the professional opinion of the doctor involved, a failure to continue the operation would seriously endanger the health of the patient.
73 See Article 8 of the Convention.
74 In the case both of a minor and of an adult without capacity to consent, the yardstick used to determine whether that person is unable to consent is the domestic law of each signatory state. As the Explanatory Report notes,

> Since the purpose of the Convention is not to introduce a single system for the whole of Europe, but to protect persons who are not able to give their consent, the reference in the text to domestic law seems necessary: it is for domestic law in each country to determine, in its own way, whether or not persons are capable of consenting to an intervention and taking account of the need to deprive persons of their capacity for autonomy only where it is necessary in their best interests.

75 See Article 6 of the Convention. See also de Bousingen, 1996, p. 2. According to the Explanatory Report, this procedure is consistent with Article 12 of the United Nations Convention on the Rights of the Child, which states that:

> States' Parties shall assure the child, who is capable of forming his or her own views the right to express those views freely in all matters affecting the child, the views of the child being given due weight in accordance with the age and maturity of the child.

76 Under Paragraph 3 of Article 6, an adult may lack capacity because of *a mental disability, a disease or for similar reasons.*

77 De Bousingen takes the phrase *as far as possible* to mean that adults should be given a say in the situation *as far as their grasp of the situation allows.*

78 See Article 7 of the Convention.

79 Once again, the Convention refers to the domestic laws of signatory states and insists that these provisions are subject to any protective conditions imposed by these laws.

80 Also relevant is Article 9, which allows for the previously expressed wishes of a patient, currently unable to express his or her views, to be taken into account.

81 See Chapter V of the Convention.

82 This General Clause operates for the totality of Chapter V. This is significant in that Article 18 (regulation of research on embryos) is included within this chapter and hence there is a further implication that the embryo is to be considered a human being. This point is discussed in detail later.

83 These criteria were largely inspired by Recommendation R (90) 3 of the Committee of Ministers on medical research on the human being.

84 According to the Explanatory Report, the words *specific consent* are to be taken to refer to consent in respect of any and every particular intervention undertaken during the course of the research procedure.

85 Again the Convention stresses that such consent may be withdrawn at any time.

86 Article 17 of the Convention.

87 The Explanatory Report suggests that this means that in the case of children, it is necessary to evaluate their views in respect of the research. The Report goes on to say

> The rule prohibiting the carrying out of research against the wish of the subject reflects concern , in research, for the autonomy and dignity of the person in all circumstances, even if the person is considered legally incapable of giving consent. This provision is also a means of guaranteeing that the burden of research is acceptable to the person at all times.

88 The Explanatory Report stresses that this does not merely mean that there are no volunteers capable of consenting to such research. Rather, recourse to persons who are unable to consent must be the only scientific possibility of achieving such effective research. In this context, reference is made to research into diseases affecting children, where children are the only appropriate subjects for such research.

89 Such conditions are listed in Article ,16 paras. I–iv and also under Article 6.

90 According to the Explanatory Report, these conditions imply, *inter alia*, that the research project is not only scientifically meritorious, but also, '… is … ethically and legally acceptable and has been given prior approval by the competent bodies'.

91 It is notable that whereas Article 16 requires a risk/benefit analysis in assessing the validity of such research, Article 17 requires the far more stringent test of minimal risk and minimal burden.

92 The Report also refers to the fact that some research (for example research performed on children) may be ongoing and may provide no actual benefit to the child who is the subject of the research while significantly improving the body of knowledge in respect of the general nature of childhood health. The example is given of research which might lead to the replacement of invasive X-ray examinations on children with ultrasonic scanning.

93 The Explanatory Report specifically states that blood transfusions are not covered by the provisions in Chapter IV.

94 Chapter VI, Articles 19 and 20 of the Convention.

95 Because of the very serious nature of such procedure, such consent must be given expressly and specifically, either in written form or before an official body.

96 Interestingly, dialysis treatment, because of the intrusion which it imposes on the quality of life of the recipient, is not deemed to have results of comparable effect to those of a kidney transplant.

97 Article 20 (1) of the Convention.

98 This tissue is defined by the Explanatory Report as that *capable of reconstituting its tissue mass and function after partial removal.* This would include bone marrow and is significant in that it can only be transplanted between genetically compatible persons, typically siblings. Thus one of the special conditions contained in Article 20 (2) is that the recipient of potentially life saving tissue be a brother or sister of the donor.

99 As is the case with research, such opposition is final and must always be observed.

100 See earlier at note 98.

101 Article 3 of the Convention.

102 Article 4 *ibid.*

103 Chapter VII Article 21 *ibid.* According to the Report, the question of patents was not discussed in respect of this article and hence it is not to be taken to apply to the patenting of biotechnological inventions.

104 This is a reaffirmation of Article 8 of the European Convention on Human Rights and Article 6 of the Convention for the Protection of the Individuals with Regard to Automatic Processing of Personal Data.

105 Chapter III Article 10 of the Convention.

106 Article 10 (3) ibid.

107 Chapter IV of the Convention.

108 Article 13 ibid. The Explanatory Report argues that

> The progress of science, in particular in knowledge of the human genome and its application has raised very positive perspectives, but also questions and even greater fears. Whilst developments in this field may lead to greater benefit for humanity, misuse of these developments may endanger not only the individual but the species itself. The ultimate fear is of intentional modification of the human genome so as to produce individuals or entire groups endowed with particular characteristics and required qualities. In Article 1, the Convention provides the answer to these fears in several ways.

109 This could include artificial insemination, *in vitro* fertilisation or any other similar process.

110 Article 14 of The Convention. The seriousness of such diseases is a matter for determination by a member state.

111 Article 11 ibid.

112 Article 12 ibid. The Explanatory Report insists that insurance companies will not be allowed to hold a predictive genetic test on which basis insurance policies might be altered or even refused.

113 Article 23 of the Convention.

114 Article 25 ibid.

115 Article 26 ibid.

116 Article 29 ibid.

117 Article 25 of the European Convention on Human Rights.

118 Article 31 of the Convention.

119 This was originally due to be ready for signature in 1998.

120 The Committee of Ministers has already given orders for the preparation of a Protocol in this area covering issues of diagnosis, gene therapy and the use of information on the genetic heritage.

121 See *The Observer*, 23 February 1997.

122 See 'Council of Europe: No to human cloning', a press release from the Council of Europe, 25 February 1997.

123 A call for a similar step was also made by the Parliamentary Assembly and the Council of Europe Committee on Science and Technology. See 'Parliamentarians call for ban on human cloning', Council of Europe press release, 11 March 1997.

124 Interestingly, on the same day, President Chirac opened the European Conference of National Ethics Committees focusing on ethical aspects of health choices.

125 The Protocol is open for signature by the 40 member states of the Council of Europe and the other states which participated in the drafting of it (Australia, Canada, USA, Japan and the Holy See) and the European Union. Only states which have actually signed the treaty are entitled to sign the protocol.

126 The original signatories were Denmark, Estonia, Finland, France, Greece, Iceland, Italy, Latvia, Luxembourg, Moldovia, Norway, Portugal, Romania, San Marino, Slovenia, Spain, Sweden, the former Yugoslav Republic of Macedonia and Turkey. Since then, Lithuania (25 March 1998) Slovakia, (31 March 1998) the Netherlands, (4 May 1998) and the Czech Republic (24 June 1998) have all signed the protocol.

127 See 'Europe takes a stand against human cloning', a Council of Europe press release, 12 January 1998.

128 On the other hand, issues relevant to cloning will be discussed later in the consideration of the problems relating to the Convention's approach to embryonic research.

129 See the Explanatory Report to the Additional Protocol, *http://www.coe.fr/oviedo/rap168e.htm.*

130 For the purposes of the Protocol and Convention, the term *genetically identical* is taken to mean *sharing with another the same nuclear gene set.* See Article 1(2) of the Additional Protocol. It should be noted that the preamble to the protocol recognises the fact that embryo splitting does occur naturally resulting in the birth of identical twins.

131 Article 1(1) of the Additional Protocol.

132 See the Explanatory Report to the Additional Protocol at p. 1.

133 See ibid. at p. 2.

134 See for example the 1988 CAHBI Report on Human Artificial Procreation, specifically Chapter VII, Principles 16–21 on Acts and Procedures carried out on Embryos. In similar vein, from 15–18 December 1996, the Council of Europe organised a Multidisciplinary Symposium on Bioethics, focusing specifically on the issues of medically-assisted procreation and the protection of the human embryo.

135 Article 18 of the Convention.

136 It should be noted that it is difficult to reconcile such vague protection with the inherently destructive nature of embryonic research.

137 It is notable that the Convention, does not seek to prevent research on embryos *in vitro* but merely prohibits *the creation* of embryos for research purposes. Thus a good deal is left up to the domestic law of signatory states in controlling the procedure of embryonic research.

138 Both Germany and Poland abstained from signing the Convention on the basis of concerns in the area of embryonic research.

139 It has been suggested that there will always be arguments in this area because of *national, cultural, philosophical and religious differences between member states*. See Roscam Abbing, 1996, p. 202.

140 Resolution No. 3 on bioethics from the 17th Conference of European Ministers of Justice, Istanbul, 5–7 June 1990.

141 *Paton v. UK*, cited earlier. See also Kingston, Whelan, and Bacik, 1997, pp. 79ff.

142 *Roe v. Wade* 410 US 113 at p. 163.

143 In fact, the US Supreme Court in *Roe*, very definitely *did* select *one theory of life*, i.e. that unborn life became worthy of protection only at the point of foetal viability, and hence that before that time abortion was permitted on demand and as a matter of right.

144 The overriding effect on US abortion law of the recognition of a constitutional right to abortion in *Roe v. Wade*, has been lessened slightly in recent years by an acceptance by an increasingly anti-abortion judiciary that there may be a compelling state interest in potential life at all stages in pregnancy. See *Webster v. Reproductive Health Services*, 492 US 490 (1990) and *Planned Parenthood of S.E. Penn. V. Casey*, 112 S.Court. 2791 (1992); 120 L. Ed 2d 674. See also Kingston, Whelan and Bacik, 1997, pp. 268–89.

145 It is arguable, however, that Ireland's decision to enshrine in its constitution a very vague unborn right to life clause was in fact based on the success of simplistic argumentation. See Hesketh, 1990.

146 See Williams, 1994, p. 73. On the other hand, it is arguable that where it is necessary to draw a line for legal purposes, then this should be done even where line drawing may appear arbitrary at any and all points. See Dyibikowski, 1975, p. 102.

147 Such protection exists under Article 40.3.3. of the Irish Constitution and also, possibly, under Sections 58 and 59 of the Offences against the Person Act, 1861.

148 See the 1996 Report of the Irish Constitution Review Group at pp. 275–6. See also Sherlock, 1989, pp. 20ff. and Brahams, 1983.

149 In fact, according to Sherlock, the likely consensus in 1983, when Article 40.3.3 was being discussed in the Irish parliament was that the 'unborn' mentioned in that article would receive protection either from fertilisation or implantation but certainly no later than this. However, a 1982 proposal to exclude from the terms of Article 40.3.3. fertilised but non-implanted embryos was rejected. See *Seanad Debates*, vol. 100, cols. 1091, 1149. See also *Dáil Debates*, vol. 339, cols. 1362 & 1365. In fact some Irish parliamentarians appeared oblivious to the consequences of protecting life from fertilisation, stating that they wished to see life protected from conception while also stating their support for postcoital contraception in cases of rape.

150 *Roe v. Wade*, 410 US 113 (1973).

151 In the United States, a distinction is made between non-fundamental rights or liberty interests, which may be set aside comparatively easily, and fundamental rights which can only be set aside if there is a compelling interest requiring this. See Tribe, 1988, chapter 5.

152 *Roe v. Wade* at p. 163. For a criticism of the use of viability as the point at which the law would start to protect unborn life. See Noonan, 1970, p. 52.

153 Thus Ronald Dworkin (1989, p. 52) comments that

> *Roe v. Wade* understood in that way did not balance a woman's rights against the competing rights of a foetus or of anyone else. Rather it identified a scheme of regulation that could meet a state's most powerful needs without compromising a woman's rights at all.

154 Kommers, 1992, ch. 11 and 1977. See also Kingston, Whelan and Bacik, 1997, pp. 308–31.

155 This was deemed to exist in Article 2(2) of the German Basic Law (or *Grundgesetz*) and was recognised in the *First Abortion Case*, a decision of the German Constitutional Court, of 25 February 1975 (1975) 39 BVerfGE 1.

156 Essentially, a woman is entitled to abortion on demand for the first 12 weeks of pregnancy if, prior to the operation, she undergoes mandatory counselling. In such circumstances the abortion is not lawful but equally is deemed non-punishable. Abortions in the first 12 weeks where conception is the result of a crime and abortions without time limit where there are medical reasons justifying such abortion are deemed lawful. This position is set out in Sections 218 and 219 of the German Penal Code as inserted by the Pregnant Woman's and Family Aid (Amendment) Act, 1995. This Act was itself a response to the *Second Abortion Case* in 1995 dealing with proposed 1990 abortion legislation for the newly reunified Germany. See Kingston, Whelan and Bacik, 1997, pp. 323ff.

157 See 'Germany: Strict Laws Protecting the Embryo', at p. 14 in *Bioethics* cited earlier.

158 A similar reaction was felt by many to the destruction, in August 1996, of 3,300 frozen embryos in England. See 'Destruction of Embryos to go ahead despite protests', *Irish Times*, 1 August 1996 and 'Vatican condemns pre-natal massacre of UK embryos', *Irish Times*, 23 July 1996.

159 For two contrasting views on the nature of rights see Glendon, 1993, cited earlier and Dworkin, 1977.

160 In *Palko v. Connecticut*, 302 US 319, Cardozo J. referred to fundamental rights as being *implicit in the concept of ordered liberty*. Similarly Scalia J. in 1989 suggested that fundamental right were those that were *so rooted in the traditions and conscience of our people as to be ranked as fundamental*. See *Michael H. v. Gerald D.*, 109 S.Court. 2333, 1989.

161 In the context of abortion law, Ronald Dworkin describes this as the distinction between a detached (i.e. non rights based) and a derivative (i.e. rights based) approach to the issue. He favours the former approach, seeing it as one which may be justified either by religious or secular logic. See Dworkin, 1993, pp. 9ff. He appreciates the arguments that a society where, for example, abortion is seen as being ethically indifferent or as commonplace as an appendectomy, would be a callous one, not because it was a rights rejecting society, but because it had lost an ethical tension in respect of an important social value. As he put it, 'Part of the sense of the sacred is a sense of taboo, and it is surely harder to maintain a taboo against abortion and to raise one's children to respect the [value of life] in a community where others not only reject it, but violate it openly…'. See ibid., p. 149. For a criticism of Dworkin's thesis see Keown, 1994, p. 671.

162 See Aristotle, 1992 and 1980. For a recent if modified endorsement of this thesis see George, 1993.

163 See Aquinas, 1991.

164 See Fitzjames Stephen, 1967.

165 See Devlin, 1965.

166 See Mill, 1991.

167 See Mill, 1991, p. 14 for the view that

> … the sole end for which mankind are warranted individually or collectively in interfering with the liberty of action of any of their number is self protection. The only purpose for which power can be rightfully exercised over any member of a civilised

community against his will is to prevent harm to others. His own good, either physical or moral, is not a sufficient warrant. He cannot rightfully be compelled to do or to forbear, because it will be better for him to do so, because it will make him happier, because in the opinion of others, to do so would be wise or even right.

168 Thus at pp. 6–7, Devlin argues that the protection of the institution of marriage, the principles that one cannot consent to an assault and that punishment of crime may be gauged on the basis of the moral nature of the crime and the criminalisation of euthanasia, suicide, attempted suicide, suicide pacts, duelling, abortion, homosexuality, incest, bestiality, bigamy and obscenity are all instances of the enforcement of morality. For criticism see Hart, 1963, pp. 25ff.

169 Lord Devlin said that he would gauge the public morality by reference to the views of the 'man in the jury box'. See Devlin at p. 90. Unfortunately he gives no clear explanation of how these views are to be discovered.

170 Basil Mitchell, 1967, pp. 16–17, interprets Devlin as meaning that only the 'essential' aspects of a society's morality could be enforced. Again however, no assistance is given as to how to distinguish essential and nonessential aspects of a society's morality.

171 For a suggested methodology for distinguishing morality and prejudice see Dworkin, 1977, pp. 248ff.

172 It is arguable for example that just as Lord Devlin's theory on the enforcement of morality allowed for the enforcement, not only of morality but also of prejudice, so also if the Council of Europe tried to enforce ethical norms in the area of biomedicine simply because of their content and without regard to the consequences of their violation, it might be difficult to separate genuine ethics from fear of progress, etc. See Butler, 1997, p. 324.

References

Aquinas, T. (1991), *Summa Theologica; a concise translation*, trans. T. McDermott, Methuen, London.

Aristotle (1980), *The Nicomachean Ethics*, trans. D. Ross, Oxford University Press, Oxford.

Aristotle (1992), *The Politics*, trans. T.A. Sinclair and T.J. Saunders, Penguin, Harmondsworth.

Bolt, R. (1960), *A Man for All Seasons*, The Drama Library.

Brahams, D. (1983), 'The Morning After Pill: Contraception or Abortion', 133 *New Law Journal*, p. 417.

Butler, D. (1997), 'Calls for Human Cloning Ban Stem From Ignorance', *Nature*, 387.

Byrne, R. and McCutcheon, J.P. (1996), *The Irish Legal System*, Butterworths, London.

Cox, N. (1997), 'Sacrilege and Sensibility: The Value of Irish Blasphemy Law', 19 *Dublin University Law Journal*, p. 87.

Devlin, P. (1965), *The Enforcement of Morals*, Oxford University Press, Oxford.

De Bousingen, D.D. (1996), 'Health, Research and Human Rights', *Forum*, Vol. 4.

Dworkin, R. (1977), *Taking Rights Seriously*, Duckworth, London.

Dworkin, R. (1989), 'The Future of Abortion', *New York Review of Books*, 28 September.

Dworkin, R. (1993), *Life's Dominion*, HarperCollins, London.

Dyibikowski, J.C. (1975), 'Lord Devlin's Morality and its Enforcement', *Proceedings of the Aristotelian Society*, p. 89.

Fitzjames Stephen, J. (1967), *Liberty, Equality, Fraternity*, reprinted Cambridge University Press, Cambridge.

George, R. (1993), *Making Men Moral*, Clarendon Press, Oxford.

Glendon, M.A. (1993), *Rights Talk, The Impoverishment of Political Discourse*, Macmillan Free Press.

Hart, H.L.A. (1963), *Law Liberty and Morality*, Oxford University Press, Oxford.

Hesketh, T. (1990), *The Second Partitioning of Ireland*, Brandsma Books, Dun Laoghaire.

Hogan, G. (1994), 'The Right to Life and the Abortion Question under the European Convention on Human Rights' in L. Heffernan (ed.), *Human Rights, a European Perspective*, Round Hall Press in association with the Irish Centre for European Law, Dublin.

Jowell, J. and Lester, A. (1988), 'Proportionality: Neither novel nor dangerous' in J. Jowell and D. Oliver, *New Directions in Judicial Review*, Stevens & Sons, London.

Keown, J. (1994), review of R. Dworkin's *Life's Dominion*, *Law Quarterly Review*, Vol. 110.

Kingston, J.,Whelan, A. and Bacik, I. (1997), *Abortion and the Law*, Sweet & Maxwell, Roundhall.

Kommers, D. (1992), *Abortion in Six Countries: A Comparative Legal Analysis* in Butler and Walbert (eds.), *Abortion, Medicine and the Law* 4th edn.

Kommers, D. (1977), 'Abortion and the Constitution: United States and West Germany', 27 *American Journal of Comparative Law*, 255.

Mathijsen, P.S.R.F. (1995), *A Guide to European Union Law*, 6th edn, Sweet & Maxwell, London.

Mill, J.S. (1991), *On Liberty and other essays* in J. Gray (ed.), Oxford University Press, Oxford.

Mitchell, B. (1967), *Law, Morality and Religion in a Secular Society*, Oxford University Press, Oxford.

Noonan, J.T. (ed.) (1970), *The Morality of Abortion*, Harvard University Press, Cambridge, Massachusetts.

Raber, J. (1997), 'European Convention on Biomedicine: Tracing it back to its Beginnings' in *Bioethics*, a Europe Infos Special Issue, Mensuel de la Commission des Episcopats de la Communauté Européenne.

Roscam Abbing, H.D.C. (1996), 'Human Rights and Medicine: A Council of Europe Convention', *European Journal of Health Law*, Vol. 3.

Sherlock, A. (1989), 'The Right to Life of the Unborn and the Irish Constitution', xxxiv *Irish Jurist* (n.s.) 13.

Tribe, L. (1988), *Abortion the Clash of Absolutes*, W.W. Norton & Co., New York/London.

Williams, G. (1994), 'The Foetus and the "Right to Life"', 43 *Cambridge Law Journal*, 71.

3 The Machine in Man

DAVID MCCONNELL

Introduction: Challenges to Science and Theology

> One effect of gene research has been to make us more aware of the unity of all life, of our rootedness in nature and of our belonging to a wider whole. The physical world is not simply the stage for human history, but an essential part of the human story. Our life is continuous with the soil, the animals, the birds and the atmosphere (Fagan, 1997).

On the 13 March 1998 I watched Father Sean Fagan discuss with Dr Ian Wilmut the cloning of Dolly the sheep. I was intrigued by his sympathetic approach, as well as his enthusiasm and warm appreciation for science. This kind of discussion encourages me to try to build my own bridges with colleagues in the Arts and Social Sciences.

These are troubled times for scientists and for theologians. Scientists, who have provided a sustained flow of knowledge which has transformed our world, are maligned because some of that knowledge has been put to evil use; theologians who have ways of distinguishing good and evil are scarcely noticed by the public. Yet these apparently opposing professions have much in common. Cardinal Daly of Armagh noted in his address at St Margaret's Church, Westminster, celebrating the 400th anniversary of Trinity College Dublin: 'There is a basic fellowship between searchers after truth'. He was identifying with the 'little people of searchers' of Samuel Beckett, in the context of an academic community which seeks truth in all its guises. So I can and do identify with my colleagues who are theologians. John Carey, Merton Professor of English at Oxford has observed that the aims of science 'seem identical with those of theology, in that they both seek to discover the truth'. The difficulty is that scientists and theologians seem destined to use different methods and for the most part they pass as ships in the night. Surveys show that most scientists are either agnostic or atheist.

The first objective of the natural scientist is to establish the facts of the natural world, including as much as possible about man, who is part of nature. The scientist seeks the truth about the natural world. Even with the caveat that the scientist can never know if he or she has the complete truth, the scientific method has been spectacularly successful in exploring, explaining and manipulating nature. If we cannot be sure that we have found out the full facts of any branch of the natural world, we can be sure that we are getting closer and closer all the time. To take one example, we have found out a lot about electricity in the last 200 years and the world around us is quite different because we have discovered, explained in a useful way, and manipulated electricity. Yet to this day physicists do not really understand why a positive charge is attracted to a negative one. We have learned a lot but much remains to puzzle us.

With all its success science has failed miserably in many respects. Among others it has failed to educate the general public about science, most people have no significant scientific knowledge, and science has more or less totally failed to influence the fundamental thinking of the modern churches. I think it is partly for the latter reason that the churches are in disarray and decline. If the churches are obviously unrealistic in their interpretation of the natural world the thoughtful parishioner tends to drift away. This has grave implications. Religion was a major source of ethics and could still be so – but I suspect that fewer thoughtful people will listen to the religious voice if the voice speaks in mystical language disconnected from the real world as science has revealed it so far.

There are theologians who are wrestling with the implications of science and there are scientists who are interested in dialogue. In this essay I would like to explain genetics, then say what this knowledge implies. There is now a perturbing case to be made that human beings are programmed, chemical, electrical machines – robots. If this is so it is hard to accommodate ethics. I will offer a glimmer of hope to those who behave as I do in the belief that humans are not automatons. There may be ways of reconciling 'free will' with determinism.

The State of Knowledge of Biology

We can distinguish some of the big discoveries in biology in the eighteenth and nineteenth century. Lavoisier, Bernard and others showed that organisms including mammals were composed of chemicals; organisms carry out chemical reactions and 'could be studied in terms of a machine'. To Bernard 'the organism represents ... what takes place in a gun factory' with division of chemical labours between the different cells and organs. Darwin and Wallace

(1859) showed that all creatures are related and had diverged over many millions of years through evolution by natural selection. Mendel (1866) discovered that factors, later called genes, controlled the structures and functions of organisms, and were passed on from parents to offspring. Golgi, Cajal and others showed that the brain and the nervous system was composed of myriads of neurones which behaved as electrical wires. By 1900 it was clear that man and all other animals are motile chemical machines partly controlled by electrical circuits. Plants are static chemical machines without electrical circuits. The design of the machines is passed on from parent to offspring by genes. Simple organisms had appeared first on the earth and more complicated ones were descended from their simple ancestors. Man is an ape with complex language.

So far as I can see there were only four great biological problems left to be solved in the twentieth century. How do genes work? How did the first biological organisms evolve from their non-biological ancestors? How does the brain work? What is the biological and mechanistic basis of language? We have answered the first, we have some good ideas about the second (which I do not have space to discuss in this essay) but we do not yet know much about the last two. As Jacob (1970) has put it, 'our present knowledge of (the nervous system) is on a par with the knowledge of heredity in the nineteenth century'. Not much has changed in the last 25 years – the brain and the biological nature of language are the last great black boxes of biology. Readers need to know that those who are thinking and writing in challenging ways about the relationship between the brain and higher mental activity, for example Philip Johnson-Laird, Francis Crick, Daniel Dennett, and Roger Penrose are being forced to speculate with little solid information about the molecular mechanics and electronics of the brain itself. Nevertheless we are beginning to find ways of opening these boxes through developments in computing, through novel instrumentation (e.g. PET scanners). The new genetic technology called 'recombinant DNA' will help us to decipher the molecular and cellular biology of the brain.

Genes

By 1925, before we knew what genes were made of, it was known that they must have three properties. They must be able to store the information which controls the appearance and functions of an organism; they must be able to replicate so that they can be passed on to the next generation; they must be able to change over evolutionary time, otherwise evolution could not have occured. In 1945, Avery, McCarty and McCleod in New York showed that genes were made of DNA, but their results were not widely accepted. All doubts were ended in 1953 when Watson and Crick discovered the structure of DNA – it was obvious from the structure how DNA could store information, replicate and evolve.

Storage of Genetic Information

DNA is a very long and very thin chemical. Molecules are known which are centimetres in length; a human cell has about three metres of DNA; if all the DNA in a person were put end to end it would stretch to the sun and back 90 times. Yet these DNA molecules are only 200 billionths of a metre in diameter.

DNA looks much like a chain with links. It is composed of four kinds of links referred to as A (Adenine), C (Cytosine), T (Thymine) and G (Guanine). A DNA chain can have millions of links. A gene is a segment of a DNA chain between precise START and STOP signals. It resembles a sentence which lies between a capital letter and a full stop. Thus a gene has a particular sequence of a few hundred 'links' of A, C, T and G at the beginning (which is equivalent to the capital letter and means START), then a specific sequence, up to one million links in length which has the information of that gene (the main information of the sentence), and finally a sequence of links which means STOP. The sequence of the links differs from one gene to another just as one sentence differs from another.

The DNA is wrapped up in packages which are called chromosomes – each chromosome has a single DNA chain. Man has about ten thousand million cells such as blood cells, liver cells, eye cells and each one has a complete set of chromosomes and a complete set of genes. Each gene is an instruction inside the cell and if it is 'switched on' or 'read' it will tell the cell to carry out certain chemical reactions. For example there are two genes for adult haemoglobin, part of the red substance which carries oxygen in the red blood cells. These two genes are switched on only in the cells which turn into red blood cells. When those genes and some others are switched on they cause those cells to contain haemoglobin and therefore allow them to carry oxygen. As another example, there is a gene for rhodopsin, which is only read in the cells of the retina at the back of the eye. Rhodopsin is a protein which absorbs light and turns it into electrical signals which we experience as sight.

All told man has about 100,000 different genes and each one causes specific proteins to be made in specific cells. There is a guiding principle of genetics – one gene, one protein, so man is composed of about 100,000 different kinds of protein. Proteins are chain-like molecules composed of 20 different kinds of links. The number of links and the order of the links in a particular protein are determined precisely by the order of the links A, T, C and G in the gene which codes for the protein.

Genes can therefore be regarded as the storehouses of information which specifies the structure of the proteins and the times and places when the proteins will be made. The proteins, which are chemicals, are the nuts and bolts, the cogs and flywheels, the pistons and cylinders, the light meters and sound detectors of the tiny cellular machines of the human factory. An eye cell is different from a blood cell because two different subsets of the 100,000 genes are active in the two kinds of cell.

Variation in Genetic Information

Humans are slightly different from each other because equivalent genes (and therefore equivalent proteins) differ slightly from one person to another. Our individuality is at root a genetic individuality, modified by the environment. For the most part this genetic individuality is a source of wonder and charm – we differ in the colour of our eyes and hair because eye and hair colour genes are quite variable. All differences between humans have some genetic origin. Some of us are genetically predisposed to being tall, others to being small and most of us to being average. Clearly the environment has a large effect on some characteristics (height, weight, colour of skin, likelihood of getting cancer, etc.) but it has no effect on others (eye colour, blood group, etc.). Sadly some genetic individuality is a source of pain, misery and death. Some people inherit defective genes, giving rise for example to muscular dystrophy, cystic fibrosis, haemophilia, etc. These defective genes cause the production of defective proteins in certain cells (respectively dystrophin in muscle cells, the CF protein in the lung, Factor VIII or clotting factor in the blood) and particular cellular machines cannot work properly. One child in 50 is born with a significant genetic defect, which means that 1,000 such children are born in Ireland every year, of which only a few can be treated satisfactorily. Haemophiliacs can be treated quite successfully, and the life expectancy for cystic fibrosis has been extended, but unfortunately there are no treatments for muscular dystrophy or for most genetic disorders. If we draw particular short straws in the genetic lottery of life our fate is quite literally set for us. It is plain for all to see that genetic information can be savagely deterministic.

There has been a lot of controversy about the extent to which genes determine human behaviour and I want to say something about this without prejudice to the more complex and theoretical arguments which I will raise later about the nature of biological determinism. The relationship between the genetics and behaviour has to be stated clearly and with full realisation of the social implications. Without subdividing behaviour, and without therefore examining different kinds of human behaviour, we can say in general that behaviour is a classical polygenic character with a significant environmental component. This means that each one of us has a set of genes which predispose us to behave in certain ways rather than others and equally importantly that our behaviour is influenced by our experience and environment. The realisation that there is a combination of genetical and environmental factors which affects our behaviour in a way which is different for every one of us carries large ethical implications.

This relationship between genes and environment must be understood. A genetic predisposition does not necessarily mean a sentence of genetic inevitability. One of us might have a set of genes which predisposes us to a particular drug addiction. Such a predisposition is neutralised if we never experience the drug. I might have genes which would help me to play the piano but these are of little significance if I am not encouraged to learn the piano. Some of the behavioural differences seen between different people have a large genetic basis, some a very small basis and some have no genetic basis. The differences in behaviour between the people of New Guinea and Greenland are almost exclusively environmental. In contrast it is clear that some psychiatric disorders such as schizophrenia and alcoholism, with the unusual associated behavioural patterns, have a strong genetic component. As we agonise over the genetic basis of behaviour, and taking into account a particular environment and set of experiences, we will come across cases in which we should be able to decide that a person probably had no choice except to behave in a particular way. In thinking constructively about how behaviour is determined and how we should try to influence behaviour for the better we should never forget two simple observations: we cannot change our genes and but can change both our education and our wider environment.

Replication of Genetic Information and Procreation

Since genes must be passed on from parent to offspring, it was obvious that they should be able to be copied, that is genes had to have a structure which allowed them to be replicated. When Watson and Crick worked out the three dimensional structure of the common form of DNA they found that it has an intrinsic duality. It exists as two chains which wind around one another in the long thin twisted structure called the double helix. Each chain is a sequence of links, the chemical units called A, C, T and G. The key to the structure is that the links form pairs; if there is an A on one strand there must be a T opposing it on the other strand. Likewise if there is a C on one strand there must be a G on the other. So the sequence on one strand has the complementary sequence on the other. A DNA molecule is not just a sequence of A, C, T and G; it is a sequence of the pairs AT, TA, CG and GC.

To get an idea of the structure of DNA, imagine a rubbery ladder with its feet stuck in the ground. Now imagine that someone twists the top of the ladder so that the sides of the ladder turn around one another, while the sides are held apart by the rigid steps. This helical structure resembles DNA. There are two kinds of step, an AT step in which an A of one side is paired with a T on the other, and a GC step in which a G is paired with a C.

The beauty of this structure is that it immediately showed to Watson and Crick how it could be replicated. If the steps could split the two complementary strands could untwist and separate. Then each separate strand could act as a template on which a new strand could be laid down. One double-stranded structure would be replicated to produce two new double-stranded structures, each with identical sequences of AT, TA, CG and GC.

Now we can describe how cells replicate themselves. Consider one chromosome. First the double-stranded DNA molecule in the chromosome is replicated giving rise to two 'daughter' DNA molecules contained in two 'daughter' chromosomes. Second the cell itself splits into two parts, each part receiving one daughter chromosome carrying a daughter DNA molecule. Human cells have 46 chromosomes. The process of making sure that each of these replicate in unison to produce 92, and that 46 are passed on to each of the two daughter cells is complex, precise and elegant. It can be seen using an ordinary microscope and can be filmed. It is called mitosis.

Simple organisms such as bacteria and amoebae are composed of single cells and so procreation is essentially a matter of the replication of cells. The procreation of animals and many plants is remarkable in that it requires the union of two germ cells derived from two parents, that is reproduction is sexual. The germ cells are produced by a special kind of cell division called meiosis which halves the chromosome number. So human sperm cells have 23 chromosomes as have human egg cells. Every chromosome in the sperm has its counterpart in the egg. When a sperm and egg combine they produce a fertilised egg with 46 chromosomes. So an ordinary human cell has 23 pairs of chromosomes, one of each pair being paternal and one maternal.

Evolution of Genes and Genetic Information

When DNA replicates it does so with great precision. So a model DNA sequence of 10 links:

ATGGCCGTACT
TACCGGCATGA

will replicate to produce two faithful copies

ATGGCCGTACT TACCGGCATGA	and	ATGGCCGTACT TACCGGCATGA.

However, replication is error-prone. Mistakes are made once in a billion times. This means that if the model sequence above was to be replicated over and over again a mistake will eventually occur. The sequence

ATGGCCGTACT
TACCGGCATGA

might replicate to give

ATGGCCGTACT TACCGGCATGA	and	ATGGCCGT**G**CT TACCGGCA**C**GA.

The change of an AT link to a GC link is called a mutation. Complex mutations may cause gross changes in the length and arrangements of the DNA sequence. If there were no mutations evolution could not have occurred.

Some mutations are advantageous and a novel gene carrying a mutation may code for a novel valuable protein which confers a selective advantage on the individual which receives that novel gene. Such a person may be resistant to a disease, survive an epidemic and pass on the gene to his or her descendants. Such a gene has been found in people who are resistant to AIDS. Millions of Africans are infected with AIDS and there seems nothing to prevent the disease from raging on and killing most of those who are afflicted. Those few who carry the resistance gene are set to be the main progenitors of the next generations of Africans.

Man, a Robot Programmed by Genetic Information?

The Danish comedian, Storm Petersen, joking about the Nobel Prize winner, Niels Bohr, said that it is always difficult to make predictions especially about the future. I will nevertheless make a judgment which has the quality of a prediction. We now know the essence of genetics and this knowledge has startling implications for our understanding of man. At face value man is a genetic robot, that is, he is a chemical machine, programmed by genetics and capable of replication and evolution. He is very closely related to the anthropoid apes which are men without complex language.

If man is a chemical machine whose structure and function is fundamentally specified by the genetic information encoded in his DNA then all his organs are just elements within that machine. Although not yet fully elucidated, there is every reason to believe that the brain is just another organ which is built according to genetic information and functions according to a genetic programme. It is increasingly obvious that the brain is a kind of computer. When it is switched on, that is, when we are awake, we are conscious. The mind may be viewed as nothing more than the brain at work, a thesis reviewed in more detail by Crick in *The Astonishing Hypothesis* and by Dennett in *Consciousness Explained*. 'The scientific belief is that our minds can be explained by the interactions of nerve cells (and other cells) and the molecules associated with them' (Crick).

The brain receives information through the senses just as a computer receives it from a keyboard. The brain can store and process the information according to programmes which are intrinsically specified by the initial genetic information and self-modifying by experience. These programmes are of course different from the usual computer programmes because they are rather fuzzy and error prone (computers are much better at maths), but in other ways the brain is much cleverer than the best computers (for example in using language, in the composition of music, in recognising faces, etc.) and the programmes are changing continuously through some unknown feedback process. The brain and its programmes are able to mature with input (learning). Unfortunately because they are built of organic molecules and not from inorganic materials such as silicon and metals, our programmes decay (with age). We are gradually learning about the genetics and biology of the brain and I expect there will be some big surprises. But there is no evidence that the brain has any properties which cannot be explained by a combination of genetics, chemistry, physics and informatics.

Ordinary computers are deterministic. A given input always gives the same output. One fact which permits us to build such computers is that the components are essentially a stable set of atoms and molecules; in functional terms a computer is the same thing from day to day. While there is a lot of motion of electrons and a lot of atomic and molecular vibration, the atoms and molecules are more or less fixed in space, and the atoms and their electrons behave in a deterministic way. Ordinary computers have been designed so that a particular instruction will always produce the same response.

Superficially the brain does not seem to be like this and people may think that brains are not computers; even if we admit that brains are a bit like computers we have difficulty in admitting that they are deterministic. However there is as yet no reason to believe that our brains are not deterministic; later I will try to find reasons why they might not be.

Let us consider the playing of a piece of piano music and for the sake of the argument let us ignore everything except the two roles of the brain in receiving the input from the music and the first level output of sending the instructions to the hands to play the piano. It is obvious that seated in front of a sheet of music a musician cannot play the music exactly the same way time after time. There will always be variation in the times between notes and the strength of the notes. If the music is long and complex, he or she will play a few wrong notes each time. It seems as if the brain has a lot of scope for free will. But I think this is not a proper deduction. One important reason for the variation is that the brain is not a stable structure. It is made of countless trillions of atoms and molecules most of which, unlike their counterparts in a computer, are in continuous motion. So the same input, the music, fed in to the brain many times, is not being seen by the same brain. The state of the brain is changing all the time.

Nevertheless we have no reason to believe that the brain is not dealing with the music in a completely deterministic way. Successive renditions of the music do vary; they can be said to differ in a way which is determined precisely and unavoidably, by the differences in the brain due in a crucial way to the fact that brain is an unstable gooey mass of molecules and atoms floating in and around watery cells.

So the brain can be viewed a wonderful computer and it is changing from second to second and so far as I can tell it is deterministic. If it is a deterministic machine, then my behaviour is deterministic, directed by a combination of an underlying deterministic genetic programme modified by environmental experiences which themselves are deterministic. How can I explain the human experiences of thought with the apparent experiences of reasoning and judgment as the output of an electrical brain? How can I conjure up free will in this Newtonian universe? Without free will where is ethics?

Two Glimmers of Hope for Non-determinism

As a geneticist I have little knowledge of neurobiology, behavioural biology, computer science, informatics or physics and I am acutely aware that I am treading on dangerous ground when I move into the territory of other sciences. But I was asked by the editor at least to present some questions which are at the boundaries of my own science and at the edges of my own comprehension. I do this with a great sense of inadequacy partly because I am not sure that they are anything like the right questions. The ideas behind my questions may appear trite especially to people who have spent their life in the field of brain research. But having read a little about brain structure and function I have not found much evidence of work which is informed by genetics. Perhaps I have a different perspective and it is close to that of Crick.

I know for sure that once there was a barren earth and that organisms first appeared about four billion years ago having evolved through a theoretically reasonable but so far unproved process of pre-biotic evolution. Man is indeed descended from apes; our closest relative living today is the chimpanzee, and we and the chimps had a common ancestor about seven million years ago. Our next closest ancestor is the gorilla; humans, chimps and gorillas had a common ancestor about 10 million years ago. And so we can go on to reconstruct the tree of life, the ancestral tree of all living things on Earth today, showing how all living things are related to each other and showing when the different branches of the family separated out from each other. Man is not just related to the chimps, he is related to chrysanthemums and butterflies, even to all yeasts, bacteria and viruses. We are related to the yeast which makes our bread and our wine.

So evolution reveals the continuum of nature, from the inanimate to the biological to the thoughtful (animals other than man surely think) and most recently to the linguistic. As Dobzhansky put it: 'Nothing makes sense in biology except in the context of evolution'. It is the process of evolution which explains most elegantly 'the unity of all life … the rootedness in nature and … our belonging to a wider whole' as referred to by Father Fagan. And it is the idea of evolution which is the strongest evidence for the extraordinary hypothesis that the brain and the mind have to be explicable in terms of the science of atoms and molecules. Ordinary science says that atoms and molecules behave in a deterministic way. I see no reason to doubt that plants or worms or bumble bees behave like robotic Newtonian machines. If there is a continuum from worms to man, at what step or in which series of steps did the Newtonian biological machine evolve into the non-deterministic animal?

So how might we get away from this ladder of determinism? I think there are two possibilities worth exploring which might in the end lead to new reflections on the relationship between the brain, thought and behaviour, between the somewhat ordered electrical processes of our brains which seem to me to be deterministic and the outputs which are our thoughts and actions which I would like to be the product of free will.

The first possibility is that the brain acts by physical processes which are themselves non-deterministic. I mention this for a variety of reasons, but mainly because it is quite a common suggestion and we must keep in mind that physics has been turning up some extraordinary ideas in this century. However, non-deterministic physical processes may not be helpful because the alternative to the deterministic may be the stochastic, that is, the truly random. If the brain is based on mechanisms which have outcomes which are essentially random, this may not be a sound basis for the qualities we associate with ordered rational argument, free will and ethical decisions. This caveat is a good reason for not exploring further the (to me) daunting questions about the nature of time, space, energy and matter, the Theory of Relativity, the curvature of space, Heisenberg's Uncertainty Principle, Bell's Theorem, parallel universes and so forth.

However, I will say a few more words at the risk of digging myself in even more deeply. It is possible that the brain functions as a kind of hybrid, with some processes random – because they are the results of random subatomic processes – and some other processes deterministic. The brain may function largely through deterministic processes which overlay a stochastic system. The brain may have complex feedback loops at the deterministic level which are programmed to edit output of a lower system which is stochastic. The stochastic processes may present alternative courses of action to 'higher' level deterministic editors which have to assess the likely outcomes and make a 'judgment' about which course of action to pursue. The deterministic processes may themselves be strongly or weakly programmed through a combination of initial genetics and later experience (education?) to 'decide' on a course of action in a way which we recognise as either ethical or unethical. However even if such a 'decision-making' system does exist and is capable of choosing to make what appear to be ethical decisions and if there is a genetic basis for this part of the system, then such a decision-making system, will have evolved under evolutionary pressure because it has been advantageous. In other words, a decision-making system which tends to make ethical decisions is not revealing free will but the biological value of what we call ethical behaviour. The fact that some people make unethical decisions from time to time may reflect variation in the structure and function of the decision-making system which is partly genetic and partly environmental.

The second glimmer of hope for a non-deterministic mind is more hopeful. It lies in the nature of language and of knowledge and the way we should explore these as parts of physics. The question is whether interactions which are mediated by language and knowledge follow the laws of physics. If they do not then thought may not be deterministic.

Physics deals primarily with time, energy and matter. The laws of physics explain the processes by which energy and matter are redistributed or transformed during time. The question in my mind is how does physics deal with the idea of information. Physics, for example through the second law of thermodynamics, does treat the concept of entropy, which is related to the concepts of order and disorder and these are related to 'information'. Can physics now move on to deal with the physical implications of complex information of the sort which has been created by man?

Consider a sentence which is read or heard. Clearly the content of the sentence is transmitted by a physical process (respectively as a stream of light or sound). The physical input signals are a sequence of photons (light) or pressure variations (sounds) and these are received by the eye or ear and converted into electrical signals for transmission to the brain where they are interpreted by systems which have been educated or entrained to make sense of them. If the brain has not been well-trained to understand the language being used, the signals will be nonsensical. So far as I am aware we do not know how the brain interprets a stream of physical signals as a sequence of words, nor do we know how the meaning of each word is 'understood', nor how the brain interprets the order and tense of the words (the syntax), nor how the brain distinguishes the appropriate meaning of a word by its context (say, distinguishing the two spoken words 'not' and 'knot'). This sort of puzzle is probably being addressed (and may have been solved) by computer scientists and informaticians who are building artificial systems which can receive and interpret normal human sentences. If an artificial deterministic computer can be built which deals with all the nuances of the transmission and processing of human knowledge, in which the computer receives and transmits instructions *in language*, with the same degrees of ambiguity and misunderstanding as is normal in exchanges between people, then we must assume that the brain is such a computer and deterministic. This is a tall order – such a computer would appear to be human. I am not sure whether such a computer can be built.

We humans have knowledge in different places, stored usually in the form of language in books, on tapes and so forth, or in our brains in some way which we do not yet understand. Knowledge can be neutral and literally meaningless unless it can be understood and put into effect. The effects of interpreting knowledge can clearly be enormous – a few people know how to release a US nuclear first strike.

Imagine for a moment that every person on earth is eliminated without any damage to the libraries or other storehouses of human information. Now the physical future of the earth will be quite different just because there is no one to use the knowledge.

How do we quantify the physical value of human knowledge (and the knowledge of other animals) and relate the measurements of different kinds of knowledge to each other (and to the physical costs and benefits of creating and using knowledge) in ways which are comparable to the relationships commonly used in physics to relate energy and mass? We need an interface between knowledge (information) and physics. It seems to me that we have to be able to create this interface if we are to find out whether knowledge expressed in language (or other symbols) is deterministic. I suspect that it is going to be very difficult to develop a physics of knowledge (and a physics of language and the sub-linguistic subconscious electrical systems in which we 'think'). I even have a suspicion and certainly a hope that language and knowledge may not be susceptible to deterministic analysis. If either or both are outside the deterministic natural sciences then the human brain may be operating outside deterministic natural science. The brain may not be deterministic because it communicates through non-deterministic language or its sub-linguistic subconscious counterpart, and because the brain depends on knowledge which is not always interpretable in a deterministic manner.

Conclusion

At a symposium in October 1998 of the European Molecular Biology

Organisation held in Lisbon speakers addressed the problem for science in conversing with the general public. Several speakers mentioned how the public are becoming uneasy as science explains one mystery after another. We seem to be close to a theory of everything not just in physics (TOE) but in biology. It is disappointing if the public are switching away from science when so many wonderful discoveries are being made on matters as varied as the evolution of man and the origin of the universe. We are on the verge of major discoveries about how the brain works and how information is stored in forms which can be turned into language (or music or mathematics). But it is understandable that man is becoming uneasy as he finds out more about the essential mechanics of his own body and mind. I hope that we may find an escape route. The best hope would seem to be that either language or knowledge (or both) is not reducible to deterministic laws of the natural sciences.

Simple biological systems, for example bacteria, behave as if they are deterministic in the Newtonian sense. But more complex organisms which display complex behaviour, in which there is much communication between the individuals, and especially when such behaviour involves even primitive 'language', may not be deterministic. Language as it becomes complex possibly does not obey Newtonian physics. The evolution of behaviour and especially of language, with all the opportunities for reflection, introspection, self-awareness, memorisation, misinterpretation, misunderstanding and opportunities for choice (if these are real) may have seen to that. Language may have allowed us a remarkable capacity to be non-Newtonian and to exercise free will. We might muse that if people believe they have free will, that in itself may be enough to cause the whole human endeavour to be non-deterministic.

In any case most of us believe that we are independently-minded, thoughtful, creative, imaginative creatures with free will. We have invented a glittering World Three (as used by Popper) which may be non-deterministic, full of great art and literature, full of great questions to which we have provided some satisfying answers, full of great inventions, and much knowledge, much of it useful and much of it dangerous.

I will close with a reference to de Chardin. I have an English translation of *Le Phenomene Humain – The Phenomenon of Man* first published by the French Jesuit biologist, Teilhard de Chardin in 1947. De Chardin was a passionate believer in evolution and believed that everything about man could and should be investigated scientifically. He had been prevented by the Jesuits from teaching and he was 'exiled' to China where he did some excellent work on fossils – in fact he was member of the expedition which unearthed Peking

Man (an example of *homo erectus*). Some of his manuscripts, left to a friend could not be published until after his death.

De Chardin was driven by reason and a search for both scientific and religious truth. He was a close friend of Julian Huxley, the eminent humanist, zoologist and populariser of the theory of evolution (author of *Evolution, the Modern Synthesis*), who was Director General of UNESCO from 1946–48.

Huxley wrote the Introduction to the English edition of *The Phenomenon of Man* (1959). He concluded:

> We, mankind, contain the possibilities of the earth's immense future, and we can realise more and more of them on condition that we increase our knowledge and our love. That, it seems to me is the distillation of 'The Phenomenon of Man'.

De Chardin for his part in his Preface says this: 'Like the meridians as they approach the poles, science, philosophy and religion are bound to converge as they draw nearer to the whole'.

We need more alliances of the kind epitomised by de Chardin and Huxley.

References

Carey, J. (ed.) (1995), *The Faber Book of Science*, Faber and Faber, London.
Crick, F. (1994), *The Astonishing Hypothesis*, Simon and Schuster, London.
Daly, Cardinal (1992), *Hermathena: A Trinity College Review*, Quatercentenary Papers.
De Chardin, T. (1959), *The Phenomenon of Man*, trans. by B. Wall, Collins, London.
Dennett, D.C. (1991), *Consciousness Explained*, Little, Brown, London.
Dobzhansky, T. (1951), *Genetics and the Origin of Species*, 3rd edn, Columbia University Press, New York.
Fagan, S.M. (1997), *Does Morality Change?*, Gill and Macmillan, Dublin.
Huxley, J.S. (1942), *Evolution, the Modern Synthesis*, Allen and Unwin, London.
Jacob, F. (1972), *Chance and Necessity*, trans. by A. Wainhouse, Collins, London.

4 Designing Life: The Role of Inheritance

DAN BRADLEY

Genetics is a science which is perceptibly on the move. Technologies now exist for manipulating the genetic material of organisms as diverse as viruses, plants and even mammals. These are impinging more and more on the public mind. The controversy surrounding recent field trials of sugar beet which have had extraneous genetic material inserted into their genomes is a case in point. The serious potential of genetic technology is reflected in an unprecedented presence on the global stock markets. For example, the agrichemical giant, Monsanto, has invested $6.6 billion in biotechnology and genomics since 1997, multiplying its stock value in the process (Enriquez, 1998). Sequencing of the entire genetic codes of a host of organisms is either finished, or is within sight of completion; including, most famously, the Human Genome Project. Molecular genetic knowledge is surfacing ubiquitously in human pursuits: from archaeology to actuarial science, from farming to pharmaceutics.

Genetics at its core seems harmless enough, it is simply the study of the biological nature of inheritance. However, partially as a result of the pace of genetic discovery and its increasingly obvious application, a credibility gap exists between practitioner and public. It is one of potentially serious consequence and, if anything, is likely to widen in the coming years. I believe that an important part of this gap and a major source of fear is an erroneous, but widely held, image of the gene and the power of genetic manipulation.

Almost exactly 100 years ago, the laws of heredity first published by Gregor Mendel, were rediscovered. The great contribution of the plant-breeding Augustinian abbot was his discovery that biological inheritance was particulate. He found that pea plants showed a number of inherited traits which were discrete and binary in their expression, for example, tall and short, wrinkled pea and smooth pea. Moreover, these segregated through the generations in a reproducible, independent fashion which seemed to be the result of the passing on of discrete and unchanged units of inheritance. These units, later coined as genes, show an immortality and inalterability that their hosts do not. This crucial separation between the messenger and the message is Mendel's legacy and is central to the subject of genetics. Well known, so called Mendelian traits in humans include the relatively trivial such as hairy ears and the ability to roll one's tongue, but also very serious examples of genetic disorders such as cystic fibrosis, which result from the presence of genes which fail to perform vital functions correctly. The effects of genes reach into almost every facet of humanity. Indeed, scarcely a month goes past without another (usually overreaching) media report that the gene for another human trait – obesity, homosexuality, intelligence – has been found.

Genes are physical realities and have been shown to reside in a linear fashion on specific cellular constituents called chromosomes. The genetic material within these has been identified as desoxyribonucleic acid, or DNA. The special quality of DNA is that it may exist as supermolecules; extremely long strings of four constituents, or bases. These bases are labelled for short, A (Adenine), G (Guanine), C (Cytosine) and T (Thymine), and through the sequence in which they occur, they write a code in the DNA. Each human possesses three billion bases of this quaternary code, the vast majority of which has no discernible function. Genes are the intelligible sentences written in what seems to be mostly a wilderness of meaningless jumble.

At the molecular level, any particular gene functions by coding for a protein. Proteins interact in the cellular environment to give us our biochemistry. They form the complex building blocks from which cell structure and activity emerge and, via the accumulation of cells into organs, they mediate physiology. The composition and hence functional activities of a given protein are enabled and limited ultimately by the sequence of the gene which codes for it.

A human has in the region of 120,000 genes and each person has a corresponding complement. However, the sentences of code, whilst equivalent in all humans, will have variations between individuals. These differences between homologues are usually only small in extent but may be large in consequence – as a change in a single letter may alter the core meaning of a phrase, or even render its syntax unintelligible. Variations in the chemistry of the protein constituents of our cells and hence organs, lead ultimately to differences in our physiological landscapes. These differences may be slight, alterations in the gradient of a biochemical curve, or may perhaps be major; peaks and troughs which profoundly effect the way in which we interact biologically with the environment.

There have been great advances in the last decade in our understanding of the genetic constitution of human beings. There is a serious effort under way, termed the Human Genome Project, to determine the sequence of all three billion units of human genetic code. Coupled with this is the intensive study of gene variations between humans in order to understand the basis of our biological differences. This, of course, is directed primarily to those differences which impinge on health. To date, towards 100 genes which display defects which result in disease have been sequenced, leading to new insights and diagnostic techniques, but only rarely as yet, therapies. However, therapy remains the strongest aim of this research which is attested to by the huge investments made in time and resources toward this.

The devastating effects of the single genes which cause disorders such as the brain-wasting of Huntington disease or the muscle deterioration of Duchenne muscular dystrophy give these predominance in human genetic research. However, strong single gene effects may provide a poor paradigm for understanding wider genetic influences on human nature. In popular discourse, the genetic map resulting from the Human Genome Project often appears as a map of fate. It sometimes seems that our sexual behaviour, intelligence, sporting prowess, even possible criminality are all subroutines in a computer programme, written before we were born and for which we function merely as passive hardware. Famously, Richard Dawkins has written extensively describing human beings as survival machines for genes. Thus, genes 'swarm in huge colonies, safe inside gigantic lumbering robots, sealed off from the outside world, communicating with it by tortuous indirect routes, manipulating it by remote control' (Dawkins, 1989).

This is one example of a variety of powerful metaphors used to explain gene action which have a single unifying theme. It is that genes are active and we, their hosts are passive. Thus biological life is an extension of the Newtonian clockwork paradigm (for example, *The Blind Watchmaker*) or, humans are mere beehives in which genes buzz around like worker bees, or alternatively, genes themselves are like a 'colony of viruses [which] … work together to produce the body and make the body do what is good for all of them' (Dawkins, 1995). There is a pervasive philosophy of genetic determinism, the idea that genetics is a sufficiently causal condition for a range of traits including even those of human behaviour and personality. In this perception genes are the masters and hosts are automatons. This is a concept (perhaps a meme!) which may be more a legacy of Calvin's sixteenth century ideas on predestination rather than Mendel's laws, as geneticist and author, Steve Jones, has cleverly pointed out (Jones, 1996).

The subject of designing life through genetics is partly a question of whether genes can and should be altered. However, it is more fundamentally a question of what altering genes can achieve. What in fact to genes do? The answer to this lies partly in the sister subject to Mendelian molecular genetics, quantitative genetics. Quantitative genetics is an old discipline. It seeks to understand inheritance by looking from the top down – a direct complement to the bottom-up approach of the molecular gene-hunter. Quantitative genetic traits are heritable traits which are not discrete. They are multifactorial, partially under genetic and partially under environmental influence. The genetic components seem typically to result from the action of many genes with something in the order of 5–10 major loci. Many very important human traits are quantitative in nature and include predispositions to diseases such as: schizophrenia, manic depression and other psychiatric disorders, diabetes, epilepsy, cardiac disease and rheumatoid arthritis. Unlike most single gene Mendelian inherited disorders, quantitative medical genetic traits affect very large proportions of the public and account for a huge amount in general health care expenditure.

Quantitative genetics also forms the basis of the most effective, by far, of genetic technologies; animal and plant breeding. In one sense, the decision on whether or not to design life has been made by our agricultural predecessors over 10,000 years ago in the Near East. The invention of farming in the Fertile Crescent brought the genetic destinies of the wild relatives of domestic plants and animals firmly under human control. The resulting genetic modification over millennia now leaves few of the species which provide us with the bulk of our food able to survive independently of human husbandry. Also, the application of scientific breeding in the present century has resulted in massive gains in productivity in many species. For example, milk production per dairy cow has doubled in Ireland within a 20-year period and average egg production per hen is now over five times what it was in the 1930s.

The highly visible success of animal breeders in nineteenth century Britain was a factor in the rise of eugenics. This was an intellectual and political movement centred on the application of breeding principles to humans for the betterment of specific populations or indeed, the human race. The movement had a prominent founder in Francis Galton, early biometrician and first cousin of Charles Darwin. It also had famous and influential adherents, such as George Bernard Shaw, Winston Churchill and Marie Stopes. Its ideas had some currency with some leading geneticists such as Fisher and Haldane. Eugenics was fuelled by a fear that the human race, the USA or the British nation (whatever) was slowly decaying because of the overbreeding of less desirable races/classes compared to elite stock. Not surprisingly, the qualities which distinguished desirable genetic types often bore a striking resemblance to those possessed by the authors of the eugenics programme themselves.

If Calvin is the spiritual forefather of genetic determinism then eugenics was a crusade, in its search for earthly perfection, against original sin. In a way it was itself an original sin of human genetics, although modern practitioners argue with justification that their pursuit is distinct. The key difference is that modern genetic medicine has a definite focus on the needs of the individual patient, rather than those of the wider population. Also, ethically, in western practice, the principle of nondirective counselling is paramount in patient-related decision-making.

The modern rejection of the eugenics movement is justified for a number of reasons. Firstly the practices recommended and ultimately utilised in some instances were a gross assault on human freedom. These included the sterilisation of the poor, the insane, the retarded; the restriction of immigration on racial grounds; attempted promotion of the birth rate of the better; and crucially, the racial hygiene practices of Hitler's Third Reich. The last was eugenic philosophy's nefarious climax and led to its decline in western societies, although it sporadically resurfaces as justification for certain government policies – most recently in China (Normile, 1998).

The second reason for the rejection of the eugenic programme for designing life is scientific. Many of the traits which its proponents wished to promote were poorly defined and of unknown or low heritability. Additionally there is little to suppose that the selection of lower caste groups owed to anything other than the received prejudices of the time. Thus in Britain, discussion centred around the merits of the lower economic classes and in the multi-ethnic USA targets were more commonly of specific racial groups.

The genetics of most common traits are complex. Within a population which exhibits some variance for a measured quantity, the components which contribute to this total variation may be estimated, The most fundamental separation is that into two main categories: the environmental variance and the variance attributable to genetic differences. Thus, simply:

VT	=	VG	+	VE
Total variance	=	Genetic variance	+	Environmental variance.

The estimation of quantitative genetic variance components is achieved by studying similarities between relatives, most famously in humans, by twin studies. Indeed, by deeper study, the components may be further subdivided thus:

$$VT = VA + VD + VI + VGE + VE.$$

Here the three genetic components which add together to give VG are VA (the additive genetic variance), VD (the dominance genetic variance) and VI (the interaction genetic variance). The environmental variance now includes VGE, a component for the interaction between genetic constitution and environment. Only a superficial treatment of a large subject is possible here but two points are worth emphasising. Firstly, no gene is an island. The effects of genetics on a trait involve the portion of the trait which may be transmitted in a heritable fashion to the next generation (VA), and the portion which depend on the genetic context of the present generation – interaction with sister alleles (VD) and interaction with other genes (VI). Additionally, there may be a nonadditive interaction with environment (VGE) as well as the catch-all term of environmental influence itself (VE) which, in humans, includes the myriad inequalities of human experience, from influences in the womb to results of personal choice. Secondly, the potential success of any breeding programme, is limited by the fraction of the total trait variance which is attributable to additive genetic effects (VA), not the proportion according to the total genetic influences (VG). The former is known as the narrow sense heritability and the latter as the broad sense heritability. Many important human medical traits show appreciable broad sense heritabilities, such as manic depression (70 per cent), rheumatoid arthritis (30 per cent), schizophrenia (45 per cent) and a range of psychological traits (40–50 per cent). Narrow sense heritabilities are more difficult to estimate and are invariably lower. For example, intelligence as measured by IQ (probably the most controversial of quantitative genetic traits) shows a broad sense value of 48 per cent but a narrow sense heritability of only 34 per cent (Devlin et al., 1997). Eugenicists anticipated a return for their efforts which would not be possible with values such as this latter estimate. It was mistakenly thought that the incidence of 'mental defects' could be reduced by 10 per cent in one generation and by 19 per cent and 26 per cent by the second and third generations respectively with the implementation of eugenic measures. Other psychological traits with similar heritabilities include, openness, neuroticism, agreeableness, extraversion and conscientiousness (Plomin et al., 1994). While it is a mistake to rely too much on data involving characters which are so difficult to define and also to argue too strongly from heritabilities (which are population-specific estimates), it seems reasonable to assume that there is a genetic contribution to personality.

The possibility for genetic design involving such traits is hampered by their nature. Firstly, low narrow-sense heritabilities indicate that selection of parents would have only a limited effect. Secondly, it should be remembered that whereas genetic effects are discernible statistically at a population level there is no guarantee of a breeding result in any particular case. Thirdly, many of the racial and/or class based differences invoked as a motivation for planned human procreation, for example IQ, are likely to be environmentally rather than genetically based. Of course, a further difficulty lies in a moral consensus (to which I firmly ascribe) that selective breeding, while effective and valuable in the farmed, is inappropriate for the farmer.

One way in which selection may be improved is if the actual genes involved in a trait are identified and tested for directly. The process of gene discovery, is well advanced for Mendelian traits, particularly those which involve severe disease. Moreover, a modern synthesis of statistical and molecular genetics is pursuing the causes of a number of important quantitative conditions. The genetic factors affecting production in farm animals and plants are well on the way to identification, and some progress has been made with medically important traits. Genes have been localised which seem to influence predisposition to coronary heart disease, hypertension, diabetes and familial Alzheimer's disease.

Indeed this synergy of the two traditional approaches promises much for future therapeutic developments. One indication of this potential is that genetic technology is becoming a serious concern for the major pharmaceutical giants for the first time. However, the greater impact of medical genetics is unlikely to involve genetic alteration of patients. Rather, more interest lies in using genetic diagnosis to infer predisposition to disease, to enable pharmaceutical intervention to be tailored to individuals with appropriate genetic backgrounds and in the potential of molecular biology for generating a new generation of pharmaceutical agents.

The molecular dissection of the common traits which preoccupy the mind of the greater part of the community does pose novel ethical dilemmas. There are implications for insurance practices and the potential for discrimination against people on the basis of genotype. Notably there is the potential for more efficient selection of parents, perhaps in the form of sperm donors, and even individuals via selective abortion or screening of preimplantation embryos. Some eugenic selection already exists with sperm banks of supposedly superior human stock catering for genetically ambitious prospective parents. However, this remains a rare practice and is unlikely to escalate with new technology. Notably, the discovery of genes which account for normal psychological and personality differences (such as intelligence) have proved elusive thus far, which is perhaps a function of the difficulty in assessing these in a biologically meaningful way. Selective abortion remains a contentious practice. It is used widely as an option following prenatal diagnosis for severe genetic disorders. Its use as a means to merely improve the odds of having a child born with attributes at one end of a normal spectrum rather than the other would certainly be reprehensible, but unfortunately such a practice is not inconceivable. Preimplantation diagnosis is much less widely used but is offered in some British clinics as a standard technique (Handyside and Delhanty, 1997). Some 11 diseases are currently diagnosed in this way and include cystic fibrosis, Tays Sachs disease, Duchenne muscular dystrophy, Huntingdon disease and b-Thallassaemia. Single cells may be removed from embryos at an early point after *in vitro* fertilisation (around the 12 cell stage) when such a disruption will still allow normal subsequent development. This cell biopsy may be used to determine the genetic constitution of the whole embryo and decision is made whether to implant based on this information. The process is quite inefficient and not totally without opposition. Ethical problems revolve around the status of the preimplantation embryo, and particularly the jettisoning of negatively diagnosed individuals. There is less antipathy than toward selection at a later stage and certainly less suffering is involved for both mother and child than in the practice of late abortion. At present there is no question of such technology being applied to more trivial genetic traits, but it must be envisaged as a future technical possibility.

The most dramatic means of designing life is via the direct alteration of the genetic code by molecular means and is now widely practised in a variety of experimental organisms. It is successful technology and has led to modified plants which yield widely consumed products and animal models for a large number of human genetic diseases. Transgenesis, or directly changing the human genetic code is possible, but experimental applications of this focus on delivering correctly functioning copies of genes to the tissues which are critically affected in severe genetic disorders. Ethically, these practices are akin to organ transplantation as there is no attempt to change the genetic constitution of the germline – the parts of the body which pass on genetic material to the next generation. There is a general consensus among scientists that germline manipulation is undesirable. However this may change with alterations pertaining to the more severe disorders. It should be noted that research leading to germline therapies would inevitably involve many inviable and discarded embryos and perhaps foetuses.

In sum, the technical potential for genetic manipulation of future generations depends on the traits which are involved. Many traits which are envisaged as being selectable – such as predispositions to common medical conditions and normal attributes such as intelligence – are under only limited genetic control. This lack of genetic determinism means that selection or genetic manipulation at the individual level will be confounded to an irreducible degree by chance and myriad influences which are classed together under environmental effects. More severe genetic disorders are Mendelian in inheritance patterns, and many are diagnosable. There is already selection against affected individuals being born using prenatal diagnoses. This has been termed by some a home-made version of eugenics (Kevles, 1995) but does differ from the discredited philosophy of human breeding in that the crucial decisions are deferred to parents rather than being enforced at higher levels of authority.

References

Dawkins, R. (1989), *The Selfish Gene*, Oxford University Press, Oxford.

Dawkins, R. (1995), 'A Survival Machine' in *The Third Culture: Beyond the Scientific Revolution*, pp. 74–95.

Devlin, B., Daniels, M. and Roeder, K. (1997), 'The Heritability of IQ', *Nature*, Vol. 388, pp. 468–71.

Enriquez, J. (1998), 'Genomics and the World's Economy', *Science*, Vol. 281, pp. 925–6.

Handyside, A.H. and Delhanty, J.D.A. (1997), 'Preimplantation Genetic Diagnosis: Strategies and Surprises', *Trends in Genetics*, Vol. 13, pp. 270–5.

Jones, S. (1996), *In the Blood: God, Genes and Destiny*, HarperCollins, London.

Kevles, D.J. (1995), *In the Name of Eugenics: Genetics and the Uses of Human Heredity*, Harvard University Press, Cambridge.

Normile, D. (1998), 'Geneticists Debate Eugenics and China's Infant Health Law', *Science*, Vol. 281, pp. 1118–9.

Plomin, R., Owen, M.J. and McGuffin, P. (1994), 'The Genetic Basis of Complex Human Behaviours', *Science*, Vol. 264, pp. 1733–9.

5 What Makes a Person a Person? The Limits and Limitations of Genetics

SHEILA GREENE

Many issues arise at the interface of genetics and psychology, enough to fill a few books, let alone a brief chapter such as this. The remarkable discoveries of modern genetics present us with knowledge and technologies which make great demands on our wisdom and flexibility. Psychology is just one of the disciplines which is involved in the struggle to come to terms with and adjust to the implications of the findings of modern genetics (see Marteau and Richards, 1996). Some of the issues which have particular relevance for psychologists relate directly to how we handle new information and new technologies and some to how we interpret these new findings – what they mean to our model of the person, to our theorising about what shapes personality and how we conceptualise personhood. It is the latter issue – the place of genetics in our understanding of human psychology – which is the focus of this chapter.

I want to discuss the way we think about the role of genetics in the shaping of human behaviours, personality and experience. I think that there are ideas abroad which overemphasise the role of genetics and which promote a false view of the extent to which genes dictate behaviour. I am certainly not implying that this is a state of affairs which all geneticists endorse. Nor do I wish to minimise the extraordinary achievements of geneticists. I *am* arguing that genes cannot be seen as the driving force behind human social and cognitive behaviour, that there are limits to what genetics can explain and that one should be aware of the limitations of theories which extend genetic explanations beyond their actual scope. Geneticists are in fact among those who are most aware of the limits and limitations of genetics and, many, I hope, would be in agreement with much of what I have to say.

In the attempt to arrive at an understanding of human psychology it is essential to take genetic factors into account. Genetics are crucially of interest to psychologists whether they wish to understand the forces which shape human behaviour at the species level or whether they wish to understand the origins of the psychological differences between humans. So, a psychologist interested in the behavioural dispositions of all humans might ask the question, 'Are all humans genetically disposed to violence or is violent behaviour a consequence of the way we structure our society and rear our infants?' These are also the kind of questions of interest to sociobiologists and evolutionary psychologists. If, on the other hand, the focus is on individual differences the question might be, 'Why do some people behave violently while others do not? Are these differences a matter of genetics?' From the genetics perspective, the latter questions are asked by those involved in the field called human behavioural genetics. As in many other areas of scholarship it is, of course, the case, that the questions one asks, how one frames the nature of the presenting problem, can mislead us in a very fundamental way. Thus, the question 'Is human behaviour dictated by genetics or by environment?' is premised on a view of heredity and environment as polar opposites, a way of thinking which we need to transcend.

I will talk first about those theories which claim to tell us about the genetic determinants of the behaviour of humans as a species. Contemporary evolutionary theorists, often called neo-Darwinists trace their origins back to the nineteenth century and the first pronouncements of Darwin and others about the role evolution has played in the shaping of not just the physical but the social characteristics of man. From the beginning, evolutionary theorists were drawn to making conclusions about the kind of psychological and behavioural characteristics which would be associated with 'the survival of the fittest', to use Herbert Spencer's term. They have assumed, for example, that intelligence and aggression would be needed to win out against inferior stock and that these psychological characteristics would be selected for, in much the same way as physical characteristics such as strength and good health are favoured. In the earlier versions of this theory the racist assumption was that the fittest were self-evidently the white races who were busy colonising the rest of the world. Not for the first or last time science was used to justify power politics and the status quo. For writers like the British polymath, Francis Galton, this also translated into views on the lack of fitness to breed of the lower classes. Galton coined the term 'eugenics' meaning 'good birth' or good beginnings and founded the Eugenics Record Office at the University of London in 1904 dedicated to 'the study of the agencies under social control that may improve or impair the racial qualities of future generations either physically or mentally' (Forrest, 1974).

Eugenic theory was taken up widely and formed the scientific justification for the state-sponsored sterilisation of people diagnosed as mentally handicapped in the USA and in some European countries and for the unbelievable horrors perpetrated by the Nazis on the Jews and other supposedly unfit races before and during the second world war. But, you may say, that has nothing to do with modern science or with the subject of this paper. Eugenic theory has been discredited scientifically and morally. Nonetheless I would argue that these matters are still highly relevant to us all. I ask the question 'What makes a person a person?' Our theories about what creates and defines us as persons create the backdrop for our behaviour towards each other and shape decision-making at both individual and societal levels. Part of what enabled the Nazis to behave towards the Jews in the way that they did was the fact that they ceased to see the Jews as persons, they ceased to see them as human – they saw them as vermin or *untermensch*. Even though I do appreciate that writers such as Wilson and Dawkins, the most famous promoters of modern social Darwinism or sociobiology, as it came to be called, in the USA and UK, would totally disassociate themselves from any connection with or support for the perpetrators of genocide, it is the case that the objectification and depersonalisation of human beings is still an intrinsic part of the thinking of modern social Darwinists or neo-Darwinists. This type of theorising, then, is dangerous, if it opens the door to attitudes towards others and ourselves which demean and diminish our status as persons and the meaning of personhood.

Modern social Darwinists have shifted their attention from the survival of the fittest persons or group to the survival of the gene. I will take as an example the work of Richard Dawkins who is probably the best known advocate for sociobiology on this side of the Atlantic. Dawkins is a zoologist, described by Sir Peter Medawar as 'one of the most brilliant of the rising generation of biologists'. A quotation from Dawkins, whose best-selling 1976 book, *The Selfish Gene*, was published in a new edition in 1989, reads as follows and illustrates the manner in which his theory objectifies persons: 'We are survival machines – robot vehicles blindly programmed to preserve the selfish molecules known as genes' (Dawkins, 1989). Later he states that humans are 'lumbering robots [housing genes] that created us, body and mind, and their preservation is the ultimate rationale for our existence'(ibid.). Dawkins is no doubt aware that most people would find his words somewhat shocking, our *amour propre* would prevent us from feeling good about being described as 'lumbering robots' or seeing the purpose of human life, personality, and culture reduced to the blind preservation of genes. But is he right, and is he justified in promoting this conceptualisation of the person?

One of his fellow sociobiologists says 'A person is only a gene's way of making another gene' (Konner, 1982). I would like you to consider that statement. Is it our selfish genes which make us the persons we are? Do we have to swallow this proposition, however unpalatable it may be, because it is demonstrably true?

There are a number of serious problems with the theoretical claims of social Darwinism and I will leave it to others better qualified than I to argue from the theological or philosophical perspective. I will give you my opinion as a psychologist, but you may be very sure that it is not the view of all psychologists: there are many psychologists who are strong adherents of selfish gene theory.

The first problem is that by the standards the sociobiologists claim to espouse, that is, the standards of positivist science, their science is suspect. Their hypotheses are not framed in a way that makes them open to disproof. They can never be wrong since selfish gene theory is so flexible, or even slippery, that any and every behaviour can be explained in some way or another. And no behaviour, animal or human, escapes their explanatory net. The existence of any widespread pattern of behaviour is taken as evidence that it must be adaptive for the genes concerned, thus no behaviour exists which can contradict the theory. This is an example of circular reasoning. It is on the grounds of its scientific standing that Stephen Jay Gould calls sociobiological accounts of the origins of human social behaviour 'Just so stories' after the tales of Rudyard Kipling (Gould, 1980) and in recent years he has been similarly scathing about the claims of evolutionary psychology (e.g. 'pure guesswork in cocktail party mode' (Gould, 1997)). You may recall Kipling's account of how the camel got its hump or how the leopard got its spots. Gould means by using the term 'Just so stories' that these theories, are in fact retrospective, post hoc accounts that rely on surface credibility and coherence to convince, rather than substantive evidence. Sociobiology cannot adduce empirical evidence about the origins of our social behaviour which is unambiguously and independently supportive of its claims since it postulates the origins of behaviour in assumed patterns of behaviour which occurred in our ancestors tens of thousands of years ago and both empirically and logically no evidence can be found to contradict this kind of theory.

In the past, evidence of acts of altruism, where animals or humans behave in a way which endangers their own genes while helping the survival of others, presented some difficulty for those evolutionists who argued that our own individual survival and reproductive fitness were priorities. But altruism can be readily explained, it is now claimed, by selfish gene theorists who assert that if altruism is widespread it must be adaptive and encoded in the genes. By their account altruism is adaptive since it prompts the individual animal to sacrifice itself for the greater good of its kin, thus ensuring the perpetuation of more genes than would survive through the preservation of its own single life. A relevant example would be the bird which squawks to warn its flock of approaching predators but in so doing attracts dangerous attention to itself. At the human level this concept of altruism lead the geneticist J.B.S. Haldane to jest that he would lay down his life for two full siblings or eight first cousins! What selfish gene theory chooses not to recognise is that through their culture and higher reasoning powers humans have developed moral systems and reasons for acting which place them in a different category from startled hens. It happens that selfish gene theorists rely often on animal examples which they generalise to humans with questionable logic and justification. Animals are not self-conscious (although some evidence of rudimentary self-consciousness appears to exist in the great apes) nor do they use abstract symbolic systems nor do they act with reference to moral precepts.

Within psychology the elaborations on sociobiology and selfish gene theory have come to be known as evolutionary psychology (Barkow, Cosmides and Tooby, 1992). These theorists concentrate on the explanation of specifically human psychological propensities and cognitive skills such as language which are seen as hard-wired into the human brain. Perhaps the best known representative of this group is Steven Pinker, author of *The Language Instinct* (1994) and *How the Mind Works* (1998). The capacities of the twentieth century mind, he argues, are still those which best enabled adaptation to the hunter-gatherer existence of our distant ancestors. Accepting that our mind is designed to perpetuate our genes, will help us to understand better the way evolution has designed us to function. Thus we will have a better understanding of who we are and what we are capable of doing.

Contemporary evolutionary psychologists attempt to buttress their claims about the scientific standing of their theory by making predictions in line with their theory. Thus, for example, Margo Wilson and Martin Daly predict that children will be abused more often by stepfathers than by biological fathers since stepfathers, being biologically unrelated, have no investment in the future of their stepchildren, or rather their genes tell them that this is so (Wilson and Daly, 1997). Ignoring entirely the fact that most stepfathers do not abuse their stepchildren and many birth fathers do, sure enough they find that disproportionate numbers of stepfathers abuse their stepchildren. Evidence for evolutionary psychology? Not at all. There are numerous alternative explanations which fit the data equally well, such as the fact that stepparents may not have developed an emotional attachment to their children, may see them as rivals for their partner's attention, etc. High levels of abuse are not found in adoptive fathers who have a sense of entitlement to their children and have typically developed deep emotional attachments to them dating from the child's infancy. Just how genes communicate their precise instructions to their robot host to behave violently or abusively to any stepchildren is thoroughly unclear, or even mysterious, since genes produce proteins not behaviour patterns.

Sociobiological and evolutionary explanations insist on a view of human nature as fixed in the genes and the capacities of mind being fixed by evolution. Thus a fatalism is promoted which discourages efforts to change. Evolutionary theories about the supposedly natural origins of behavioural differences between men and women or the inevitability of competition and aggression tell us that these are the facts, just accept them. They tend to be inherently conservative and supportive of the status quo.

Furthermore, such explanations, if and when they convince, distract from alternative explanations which may offer perspectives on human nature which are both more adequate and more capable of providing a basis for human morality and human freedom to change. The attraction of the biologistic and deterministic theories promoted by people like Dawkins and Pinker lies in a large part in their simplicity and in their claim to being 'scientific'.

Sociobiology, evolutionary theory and evolutionary psychology all adhere to a tradition within the natural sciences which has in general served it well, that is reductionism. Reductionism takes a number of different forms but basically entails breaking down a complex phenomenon into its assumed parts or explaining a complex phenomenon in terms of principles derived from a different and ostensibly more fundamental level of discourse – so for example saying that love between two people is a matter of compatible biochemistry. Reductionism is a most useful strategy for the biological as well as the physical sciences. But, it is not the most appropriate strategy when faced with the challenge of understanding mind or the behaviour and experience of persons. One might agree with Herbert Muller when he suggests that 'To say that man is made of certain chemical elements is a satisfactory description only for those who intend to use him as a fertilizer'!

Breaking down a complex entity into its components or elements cannot tell us much about the functioning of the intact entity, particularly when that entity is a living organism. It is certainly true, as the old aphorism says, that the whole is more than the sum of its parts. The whole is the result of the organisation and interaction of parts and moreover derives its function and its significance from its context, its relationship with its setting or environment. This is true for a bicycle which would make little sense presented as a pile of components to a South American forest dweller and it is even more true of a living organism. If we knew all there is to know about the biochemical and physiological functioning of the human brain, enough to allow us to describe the functioning of individual human brains, we would still be unable to explain or predict the psychological experiences and social responses of individual persons. The perception of the colour blue or the feeling of being sad are phenomena which are irreducible, how much more so phenomena such as pride or work satisfaction or political affiliation that are inextricably entwined with our cultural and social systems and are profoundly relational. A description of their physical correlates does not explain these psychological phenomena, all it can do is 'explain them away'. We require a theory of persons which recognises that when we are dealing with the complexities of human experience and behaviour, we are dealing with emergent phenomena which have structural and functional characteristics of a qualitatively different order from those associated with physical entities or processes.

Despite the fact that reductionism is fated not to provide an understanding of mind, self and other related constructs and processes, it is the case that, for most of its history, psychology as a discipline has been wedded to reductionism in so far as it has been wedded to the slavish emulation of the natural sciences. Only a reductionist science could have embraced behaviourism as warmly as psychology did for the middle decades of this century. Behaviourists saw humans as black boxes, the responses of this human box being shaped by environmental contingencies: they denied any meaning to mental states and they denied the usefulness of concepts like personality. Behaviourism has been replaced now as the dominant theoretical perspective in psychology by cognitivism, where a favoured paradigm is that which sees cognition in terms of information processing and likens the working of the mind to the software of the computer. It should be said that environmental reductionism is no more satisfactory than biological reductionism. Behaviourists, like Skinner and Watson, were environmental reductionists par excellence, and they were to a great extent a theoretical cul de sac for psychology.

Psychology continues to espouse reductionistic theories, as can be seen in the popularity of evolutionary psychology. Evolutionary psychologists have in recent years staked their claim as the rising universal paradigm for psychology. To evolutionary psychologists culture is an afterthought, a veneer which may change our behaviour in important ways but which, when the chips are down, is simply cloaking or confining the real impulses and motivations of the human animal – or 'lumbering robot'. I submit that not only is this demeaning of human beings, it is a fundamental misconstruction of human psychology. As Paula Derry (1996) says,

> Humans as far as we know have always had a culture … (and) are dependent upon experience with co-specifics to organise the cortex. Culture is not simply a set of constraints on what drives or emotions can be expressed but is itself a given of human biology.

So far from being merely a veneer, it can be argued that culture plays a constitutive role in forming our minds and behaviour.

This is not an argument for social determinism. Some psychological theories, in opposition to the extreme biologically deterministic theories, steer too far in the other direction. For example, contemporary social psychology is strongly influenced by a relatively new theoretical movement called social constructionism, which has much to offer, but is also in some of its expressions an example of environmental or social reductionism: that is, it sees individual personality – or as it would more typically put it – subjectivity, as a product of intersecting social discourses. Biology doesn't enter the picture nor is there much room for the person as agent. Models of personhood cannot afford to neglect either the biological or social but instead need to rethink the indissoluble relationship between these two primary forces. For example, we need to expand our definition of the biological to embrace the complexity of human thought and experience, social and cultural expressions and structures. Biology is about life and life as humans live it is intrinsically social. We would not be human if it were not for our extraordinary learning capacity, our willingness and need to learn from others, our complete embeddedness in culture, whether we live in a tribe in the Amazon Basin or in heart of a modern metropolis. Our lives are also intrinsically biological and the biology of the human being entails the capacity to learn, to use language, to use our skills and accumulated knowledge to construct mud huts, skyscrapers, poetry and scientific theories.

What we find, then, in the writings of sociobologists and evolutionary psychologists, is a strong form of genetic determinism and genetic reductionism. I would suggest that this is not the kind of theory which is useful or tenable. We need to discuss the origins of human personality and behaviour in terms which do full justice to its complexity and which do not reduce human psychology to the terms of a different and allegedly more privileged level of explanation and which do not reduce human persons to objects. The situation is not all black as far as psychology is concerned. Not all psychologists have been content with psychology's reductive orientation. Humanistic psychology, for example, which is associated with the work of psychologists like Maslow and Rogers and May, has kicked against reductionism within psychology, but it has to be said that in academia its status is low, its journals are not amongst those listed as prestigious in research assessment exercises, the professional societies do not insist that it is taught to undergraduates. One of the main tenets of humanistic psychology is that people should be understood as persons not as machines. Other movements which are gaining ground, although still at the periphery of the discipline, are critical psychology and experiential psychology (e.g. Sarason, 1981; Stevens, 1996). Critical psychology has produced in recent years analyses and, often, condemnations of the basic epistemological and political assumptions underpinning traditional psychology. Experiential psychology builds on philosophical traditions such as phenomenology and existentialism to provide a psychology which is reflexive and fundamentally respectful of the person and his or her experience of self and the world.

If reductive thinking is permitted to form the theoretical underpinning for the sciences and for the actions of the scientists now empowered by knowledge and technology to redesign life, I think we have a major cause for concern. Have no doubt about it, the advocates of sociobiology see their theory as the basis for social engineering, the redesign of human nature. In their book *Promethean Fire* (1983), Lumsden and Wilson state, 'One result of a strong human science might be the creation of a sophisticated form of social engineering, one that touches the deepest level of human motivation and moral reasoning'. Sociobiologists represent perhaps the most blatant protagonists of biological determinism but they have many like-minded bedfellows amongst biologists and psychologists. If the public accepts these neo-Darwinist theories as an accurate picture of who we are and why we behave as we do, people may be lulled into a 'scientific' way of thinking about the person which is both inaccurate and dehumanising. It is inaccurate because it is simplistic and fails to recognise the complexity and indeterminacy of human existence. It is dehumanising because of the equation of humans with machines.

The other aspect of genetics and the person which I wish to discuss is human behavioural genetics, research that has been carried out by geneticists and others on the extent to which a person's behaviour is influenced by their unique pattern of genes, or, more specifically, the role of genetics in determining individual differences in behaviour. Much work has been carried out which attempts to calculate to what extent certain psychiatric conditions or psychological capacities are determined by either heredity or environment. The results have been far from clear-cut and are not easy to interpret. One of the problems is that the methods which we have at our disposal to calculate the relative contribution of heredity and environment tell us about differences or similarities between people or about sources of variance in a population. They do not tell us how much of your or my intelligence is due to heredity or environment, they cannot. Psychologists like Robert Plomin have now claimed to find a gene marker associated with high IQ, the as yet unpublished results of a project outlined by him some years ago which , theoretically, could provide the first direct evidence of genetic determination of intelligence (Plomin, 1994). In fact, given the likelihood that there are many genes involved in providing the physical underpinning for high IQ and that these genes are likely to be different in different populations, and may interact with each other in different and unpredictable ways and with different environments in different ways, the quest for high IQ genes is not destined to produce clear results. When we are dealing with characteristics which we know are not the result of simply one faulty gene the picture becomes extremely complex. The widely cited heritability statistic (expressed as ranging from zero to one, where zero is not heritable and one is totally heritable) tells us nothing about how much of any one person's behaviour is genetic and how much due to nurture. It tells us only about the sources of variance in a particular population at a particular time under certain conditions. Taking the example of IQ, it is an ironic fact that if blacks and whites in the USA experienced exactly the same beneficial, or disadvantageous, environmental conditions the heritability of IQ would become higher, since the effects of genetically-determined individual differences, however minimal, would be the only source of variance. Heritability statistics would tell us that the heritability of black hair amongst Chinese people is zero. This is palpably nonsense if we incorrectly take heritability as an index of whether or not a characteristic is biologically determined for we know that natural hair colour is determined genetically. The heritability index of zero arises, of course, because all Chinese people have black hair and there is therefore no variance (Gottlieb, 1993).

It is interesting that most non-psychologists seem to see psychologists as the natural advocates of an environmental standpoint as far as the source of individual differences is concerned. This is not the case. Many psychologists would emphasise genetics rather than environment. As I have indicated there is a strong grouping within psychology of evolutionary psychologists and others who are biological determinists and the work of behavioural geneticists is also very influential. Many researchers and theorists within my own subdiscipline of psychology – developmental psychology – believe that the primary determinant of psychological differences between people is their genes. Many would agree with Sandra Scarr who said 'We are products of the co-operative effects of the nature/nurture team, *directed by the genetic quarterback*' (Scarr and McCartney, 1983).

As the quote from Sandra Scarr indicates we are still inclined to think in dualistic terms about nature and nurture. We think of nature or nurture or nature plus nurture whereas in human behavioural genetics what we need is to move on to a view of the person which fully incorporates the idea of *reciprocal determinism*, or as Gilbert Gottlieb puts it, *coaction* (Gottlieb, 1992). Genes are influenced by other genes and by their cellular context, they cannot stand alone nor are they impervious to influence. As Gottlieb puts it, genes are part of 'a fully interrelated, fully coactional system in which the activity of the genes themselves can be affected through the cytoplasm of the cell by events originating at any other level of the system, including the external environment'. This picture of a system in which influence is bidirectional is congruent with according psychological processes causative status rather than relegating them to the status of mere epiphenomena. It is also the case that given the complexity of the interactions involved and the capacity of novel interactions to produce novel emergent phenomena, that outcomes of gene-environment interactions and transactions are synergistic and unpredictable. The psychologist, Donald Hebb, stated that in fact behaviour is determined 100 per cent by heredity – and 100 per cent by environment, his way of emphasising how inextricably enmeshed one is with the other (Hebb, 1980). The old song claimed that love and marriage went together like a horse and carriage. Well, we all know that that isn't true, but at the level of the person heredity and environment are inseparable and you can't have one without the other. As Steven Rose argues in his recent book, *Lifelines* (1997), 'the phenomena of life are always and inexorably simultaneously about nature *and* nurture, and the phenomena of human existence are always simultaneously biological *and* social'.

A further problem with the conceptualisation of persons as the product of nature plus nurture is that it leaves no place for the person. Theories which see people as psychologically the products of sets of determinants and as *nothing but* the product of those determinants leave out of account the role of the person in determining their own development and behaviour. The definition of personhood is problematic and many different definitions are possible. However, I would suggest that two key characteristics of personhood are the subjective sense of the self as a unique and experiencing individual and the capacity for personal agency. As Susan Oyama states, 'Our conceptions of development make it easier to see people as objects formed and moved by causes than as experiencing subjects who may act for reasons – that is as persons' (Oyama, 1993), I consider that the phrase 'experiencing subjects who act for reasons' pretty well encapsulates what it is which defines and characterises personhood. Persons are sentient and self-aware, they have an active role in determining the course of their lives. But it is an important aspect of the achievement of personhood that others are willing to make that ascription, that we are recognised as possessing certain intrinsic qualities which, to my mind, imply and demand respect and other inalienable moral responses.

So when we lie in bed at night and plan what we are going to do the next day, are our thoughts simply prompted by the whispering of the genes within, as Barash puts it, or are we a part of the process of constructing our lives, actively involved and playing a part in the making of the self? Theories which are focused solely on causes and determinants which act on a passive body or mentality fail to take into account the self-making character of persons. If we are simply the product of our genes we cannot be seen as responsible for our actions. If we are merely the robot carriers of our genes we are not important as unique persons. In fact I don't think most geneticists or psychologists who are devoted to behavioural genetics actually think this way in their personal lives but their theory pushes them into this way of thinking as scientists. We have to find a way for scientists, whether they be geneticists or psychologists, to recognise and incorporate human personhood as we know it into their research and theory.

Within academia it is the humanities rather than the sciences which still cherish and cultivate the view of humans as subjects and agents, active in the creation of the patterns of their own lives, to be valued as individual and unique persons. It is time for the life sciences, whose work touches on the nature of human existence to take as their focus human subjects not human objects.

The developmental psychologist Richard Lerner asks why sociobiology and human behaviour genetics flourish when their conceptualisation of genes and of nature as separable from nurture is known by so many scientists to be 'counterfactual'. And he adds 'These questions are not merely academic. Science and public policy are, at this writing, being influenced by biologically reductionistic ideology'(Lerner, 1993). At best it might be concluded that biological reductionism appeals because of the credibility of its claim to offer simple answers to difficult questions and it is also probable that lay-people are inclined to believe forceful assertions from high-prestige scientists. The neuroscientist Steven Rose is one of those who is concerned about the grandiose claims made by some of his colleagues. He remarks on the 'gung-ho triumphalism among geneticists' which he sees as arising from the undoubted successes of the discipline but which has lead some geneticist to believe 'that their science can explain everything that is to be explained about the human condition and, indeed [can] rebuild humanity in an improved image if allowed' (Rose, 1997).

Genetic discoveries have already given us insight into the causes of some hereditary diseases and have improved our understanding of the mechanisms underpinning heredity. There is no doubt that these discoveries will continue to have an impact on our management of plant, animal and human life. With great opportunities for good will come great opportunities for evil and recklessness. If we are to manipulate human life we need to refine our ideas about what it is that we are manipulating. Naïve, brutal or misguided theories which misinterpret or overextend the reach of genetics are not a good foundation for making decisions. The challenge for the life sciences and the human sciences is to develop and integrate into their theory and research a model of the person which reflects and respects the complexity of human life and experience and which sees personhood as a reality not as an inconvenient fiction. In the light of this demand reductive, mechanistic accounts of human nature, whether they favour genetic or environmental determinants, are misleading, inadequate and potentially dangerous.

References

Barash, D. (1979), *The Whisperings Within*, Harper and Row, New York.

Barkow, J., Cosmides, L. and Tooby, J. (1992) (eds.), *The Adapted Mind: Evolutionary Psychology and the Generation of Culture*, Oxford University Press, New York.

Dawkins, R. (1989), *The Selfish Gene*, 2nd edn, Oxford University Press, Oxford.

Derry, P. (1996), 'Buss and Sexual Selection: The issue of culture', *American Psychologist*, 51, pp. 159–60.

Forrest, D.W. (1974), *Francis Galton: The life and work of a Victorian genius*, Elek Books, London.

Galton, F. (1909), *Essays in Eugenics*, The Eugenics Society, London.

Gottlieb, G. (1992), *Individual Development and Evolution: The genesis of novel behaviour*, Oxford University Press, Oxford.

Gould, S.J. (1980), 'Sociobiology and the Theory of Natural Selection' in G.W. Barlow and J. Silverberg (eds), *Sociobiology: Beyond nature/nurture*, Westview Press, Boulder, Colorado.

Gould, S.J. (1997), 'Evolutionary Psychology: An Exchange', *New York Review of Books*, 9 October, XLIV, pp. 53–8.

Hebb, D.O. (1980), *Essay on Mind*, Erlbaum, Hillsdale, New Jersey.

Konner, M. (1982), *The Tangled Wing*, Holt, Rinehart and Winston, New York.

Lerner, R. (1993), 'The Demise of the Nature-nurture Dichotomy', *Human Development*, 36, pp. 119–24.

Lumsden, C.J. and Wilson, E.O. (1983), *Promethean Fire: Reflections on the Origins of Mind*, Harvard University Press, London.

Marteau, T. and Richards, M. (1995), *The Troubled Helix: Social and psychological implications of the new genetics*, Cambridge University Press, Cambridge.

Oyama, S. (1993), 'How Shall I Name Thee?: The construction of natural selves', *Theory and Psychology*, 3, pp. 471–96.

Pinker, S. (1994), *The Language Instinct*, Penguin, Harmondsworth.

Pinker, S. (1998), *How the Mind Works*, Penguin, Harmondsworth.

Plomin, R. (1994), *Genetics and Experience: The interplay between nature and nurture*, Sage, London.

Sarason, S. (1981), *Psychology Misdirected*, Free Press, New York.

Scarr, S. and McCartney, K. (1983), 'How People Make Their own Environments: A theory of genotype-environment effects', *Child Development*, 54, pp. 424–35.

Stevens, R. (ed.) (1996), *Understanding the Self*, Sage, London.

Wilson, E.O. (1975), *Sociobiology: The new synthesis*, Harvard University Press, Cambridge, Massachusetts.

Wilson, M. and Daly, M. (1997) cited in 'Characterizing Human Psychological Adaptations', *The Psychologist*, January, pp. 4–5.

6 Crisis Pregnancies in the Age of Human Genetic Diagnosis: Women's Right to Self-determined Pregnancies and the Right of the 'Other'

HELGA KUHLMANN

In the title of this contribution, two 'rights' are joined with the conjunction 'and' which do not go together in the case of an unwanted pregnancy and which in this situation could only be joined with an 'or'. The question of crisis pregnancies has provoked and still provokes the harshest of controversies. This has been and still is a matter where different views oppose each other fiercely. Since the 1970s when the women of the new women's movement demanded the right to abortion and the abolition of Paragraph 218 in Germany, this debate has been topical and seems even to have increased in alacrity. Since then, in many Western countries abortion law has been liberalised and improved medical methods have decreased the physical risks for women. Against these legal reform groups who call themselves 'protectors of life' have defended the foetus's right to life with appeal to the 'sanctity of life'.

At the same time, the scope of decisions which pregnant women have to take has increased greatly since the 1970s. To a much higher degree than 25 years ago, pregnant women are confronted with the question whether they want to make use of prenatal diagnosis and whether they would be able to accept a disabled child, if the situation arose. The moral and legal problems are likely to become even more accentuated in the near future. At the very latest this will come true when the Human Genome Project will have been completed and all the DNA sequences will have been decoded. Then, it might be possible for prenatal genetic diagnosis to predict not only certain genetically caused disabilities and disorders but also dispositions and properties. The pressure on pregnant women to use prenatal diagnosis could mount even further; on the other hand, the possible consequences of the decision not to do so, increase as well. Already now improvements in prenatal diagnosis and even in prenatal therapy become known at ever shorter intervals. Yet, the increase in knowledge and in technical possibilities could lead to conclusions of which it is doubtful if they would be of benefit to the unborn life, the pregnant woman and future generations.

For the past 15 years, the possibility of extracorporeal fertilisation has existed for infertile couples which can be combined with genetic diagnosis. In Germany the law does not allow embryos fertilised *in vitro* to be tested for hereditary genetic disorders, and to refrain from transfer in the case of a positive result, but it is medically possible. In other countries this is a regular practice.

I would like to treat this complex topic in three steps: I shall begin by outlining how the development of genetic diagnosis and therapy has changed the experience of pregnancy in recent years. Subsequently, I shall deal with the tension present in the title. I shall contrast the right to a self-determined pregnancy with the 'right of the other'. Here, I shall refer to arguments of the German-speaking feminist and church debates. In my third part, the question will be what contribution the Christian churches and theology can make to giving greater scope to women's self-determination in crisis pregnancies and at the same time protect embryos *in vitro* from institutional and economical instrumentalisation.[1]

Changes in Pregnancy Through Methods of Human Genetics

Prenatal Diagnosis

In the last 25 years the experience of pregnancy has undergone fundamental changes through various prenatal examinations and the possibility of extracorporeal fertilisation. Pregnancy has been visualised and medicalised. For the pregnant woman, the nine months have become a time of attentive and anxious self-observation in which she is in regular contact with a physician and learns to regard herself as a kind of ecological system for the child growing in her who is presented to her in an objectified way through medical data. Her own body sense which makes her aware by and by that hidden in her body an other who is both different and her own is growing, no longer marks her experience primarily. Consciousness of risks has grown. At the same time early diagnosis and methods of prenatal therapy can help to lessen or cure a number of disorders and illnesses already in the womb or immediately after birth.

The prenatal check-ups which use among others human genetic techniques primarily are meant to monitor a risk, the risk of having a child with disabilities or genetic disorders. Apart from the three ultrasound scans which are offered to all pregnant women in weeks 8, 18 and 28, women from families with inherited genetic conditions, with diabetes, high blood pressure and other inheritable illnesses as well as women from around the age of 30 onwards can undergo the triple test. This is a blood test for finding out whether the child is likely to be born with Down's Syndrome resp. trisomy 21. This test has a high rate of uncertainty. A positive result can be further investigated in subsequent tests; normally, a sampling of the amniotic fluid (amniocentesis) is used which can be practised at the very earliest from the fourteenth week. In the tenth week already chorionic villus sampling (CVS) can be applied. In both, genetic techniques are used; with both, the risk of miscarriage is higher than the average positive result.[2]

All in all, genetic anomalies of the foetus are found in two to three per cent of women. In these cases a medical indication for pregnancy termination is issued which legalises the termination that is otherwise unlawful in Germany.[3] The medical indication has no time limit.[4]

Since most of the test results are only available after the sixteenth week, a termination through the suction method is no longer possible.[5] Therefore, with terminations after the sixteenth week, the birth process is induced.

Later terminations, however, have to be judged as very problematic. On the one hand it has become medically possible that pre-term babies already survive in the twenty-fourth week; on the other hand it is lawful to abort genetically abnormal foetuses after a medical indication until just before birth. In some clinics with late terminations of pregnancy the foetuses who could potentially survive are killed before birth with a heart injection. In order to protect gynaecologists, medical staff and the pregnant woman from such situations, the Association of Gynaecologists has decided to accept a self-obligation not to practice any terminations after the twentieth to twenty-second weeks.[6]

It can be summarised that with their decision in favour of prenatal diagnosis pregnant women are faced with a complex of consequences which they are often not aware of at that point in time. This includes the decision on the risk of miscarriage and the risk of injuring the foetus. Often pregnant women are only confronted with the possibility of prenatal genetic diagnosis in the consulting rooms of obstetricians, and many make use of them without having thought about how they would be able to cope with a positive test result. With women over 30 years of age the effect is frequently that they only accept their pregnancy after a negative result of the second ultrasound scan, the triple test (or, in addition, of amniocentesis). Only then can they look forward to their child and develop emotional ties to the unborn.

In contrast, after this period of anxiety, women with negative results often experience the feeling of a medically confirmed security that they will have a 'healthy' child. If they have availed of prenatal diagnosis on the basis of a conscious decision not to be able to live with a disabled child, they are all the more devastated if a disabled child is born after all. Only a few of them know that only a third to a half of mental and physical disabilities are caused genetically and that a large part of congenital abnormalities have other causes, e.g. an acute and chronic lack of nutrition through placental insufficiency.

Preimplantation Diagnosis

More than prenatal diagnosis, *in vitro* fertilisation has been at the centre of public debate in the past years. It is a method of extracorporeal fertilisation used since the early 1980s which is to make pregnancies possible for infertile couples.

The most recent development is to test fertilised embryos even before their implantation into the womb for genetic abnormalities. In Germany this 'preimplantation diagnosis' in the early embryonic stage of 'totipotent cells' (up to its division into eight cells) which could each develop into an individual person, as well as germline therapy are forbidden. In the United States and in a number of European countries this is permitted.

The argument given for introducing preimplantation diagnosis is that a later abortion is avoided if genetically abnormal embryos are not implanted. In the United States it is being discussed whether this method (IVF) should not only be permitted for infertile couples, but also for couples genetically proved to be at risk who can test before embryo transfer if the embryo is genetically normal. These couples can also use donor sperm to reduce the risk of genetic abnormalities even further.

Another link between IVF and human genetics consists in the fact that surplus embryos are produced through *in vitro* fertilisation. In Germany only three embryos can be created, and all of them have to be implanted. Internationally, the German Embryo Protection Act has no bearing, and embryos are the raw material for genetic research. I want to conclude this primarily descriptive section with two indications of problems.

1 Prenatal diagnosis and preimplantation diagnosis are diagnostic methods.

According to principles of medical ethics, medical diagnosis is to serve therapy. Prenatal genetic therapy of severe genetically caused disorders, however, is not legally admitted in Germany up to now and is not possible in the United States. Instead, other 'conventional' prenatal therapies have been successful in an increasing number of mild and severer illnesses.[7] The question is whether under these circumstances the aim of acquiring knowledge on the genetic make-up of the unborn justifies the risks of the diagnosis.

2 An intervention into the genetic material would also affect the subsequent generations. The consequences of genetic measures can only be known in the long term. Also germline interventions which were to serve the therapy of most severe disabilities would have to be justified to the descendants of patients who underwent prenatal therapy.

The Right to a Self-determined Pregnancy and the Right of the 'Other'

Both in the Jewish and Christian traditions and in Western culture the right of women to a self-determined pregnancy has not been a matter of course. According to 1 Timothy 2:15 the bearing of children counts as punishment and compensation for women's sin, since women were regarded as those who had given in to the temptation of the serpent and thus had brought sin into the world.

In the context of the second women's movement in the 1970s feminists in great campaigns, among others in publishing their self-incrimination 'I have aborted' in the magazine *Stern*, demanded the right to reproductive autonomy which encompassed unrestricted self-determination on the decision to have an abortion. Above all the feminist slogan, 'My belly belongs to me', was a target of criticism. By now the feminist position can no longer be seen as one, and also the advocates of reproductive autonomy argue in a more differentiated way. Two opposing feminist views are being voiced: on the one hand the demand for reproductive autonomy including the use of the new possibilities of reproductive technology and prenatal diagnosis controlled by women, as represented by Silvia Kontos in 1986; on the other hand, the demand for foregoing prenatal diagnosis, as it is put forward by the Network against Selection in Prenatal Diagnosis (an organisation representing the interests of disabled women). I shall illustrate both positions briefly.

Silvia Kontos defines the term 'reproductive autonomy' as

> autonomy in the whole process of reproduction, beginning with how one deals with one's own sexual feelings and needs, with menstruation, contraception and abortion, the shaping of pregnancy and birth up to the different possibilities of organising the care and education of babies and small children, or the transition of children to the status of adults. Autonomy means the struggle for the greatest possible scope for decision and action of women (Kontos, 1986, p. 139).

In this context, Kontos (1986) criticises the earlier women's movement who by using women's fear of a disabled child

> themselves reinforced the process of illness and disability which leads to forced sterilisation and forced abortions as a final consequence (ibid., p.141) ... and has pushed the positive aspects of motherhood aside (ibid., p. 145).[8]

She reproaches the church, the medical body and jurisdiction with using the appeal to the 'rights of unborn life over against the mother' as an instrument of social control against women. This is also where she places responsibility for the 'almost hysterical fear of a malformed child' (ibid., p. 146) which is generated through the concept of medical risk. A way out according to her is that women determine the whole process of reproduction and then make use of it including the new techniques.

Against these demands for women's reproductive autonomy, the disabled feminists and women of the 'Network against Selection in Prenatal Diagnosis' hold that antenatal tests are not directed towards therapy, but towards the diagnosis of incurable disabilities and disorders. Since in the case of positive test results the medical indication is issued after which most women have their pregnancies terminated, these tests did indeed serve for selecting disabled and severely ill persons. This is why they were to be seen in a historical context with the eugenic measures during National Socialism. Even if these tests were not to be discontinued, at least they were to be excluded from the list of medical measures covered by health insurance companies.

The handicapped lawyer and feminist Theresia Degener points out that the concept of reproductive autonomy has not only been used by feminists in recent years, but has also found its way into the advertising of companies offering prenatal diagnosis and reproductive technologies as well as of US sperm banks (Degener and Köpsell, 1992, pp. 107–21). IVF and prenatal diagnosis, sperm banks and a number of physicians and human geneticists promise women greater reproductive autonomy, yet submit them to severe criteria of selection, to lengthy physically damaging procedures and to great moral pressure if pregnant women forego these procedures.

Instead of a right to reproductive autonomy, the recognition of infertility as an illness and its possible 'therapy' through *in vitro* fertilisation seems to have moved into reach the right to have a child, and through prenatal diagnosis combined with the possibility of a medical indication even the right to have a non-handicapped child. Several times since 1982 physicians have received court sentences to pay compensation as high as the upkeep for a disabled child for not having informed parents sufficiently on prenatal diagnosis. Degener interprets this as saying that legally, to accept disability is an unreasonable demand (*unzumutbar*) (ibid., p. 68).[9]

For her, the position of the 'women/crippled women from the emancipated disabled persons' movement' comes down to demanding the 'right to abortion for unwanted pregnancies' while equally criticising eugenically selective abortions (ibid., p. 105). She also reproaches the pro-life groups of functionalising disabled people and their criticism of the eugenic indication for their campaigns without engaging themselves actively for a friendly and accommodating world for disabled people.

While both positions promote reproductive autonomy for women and by implication question an unrestricted right of the embryo, they dissent on two issues: one, the evaluation of whether women will be able to use the new reproductive technologies for themselves, or whether they will invariably become an object of science, technology and economy; second, the fundamental question whether women's right to reproductive autonomy includes the right as far as possible not to have a disabled child. This is a right that the churches deny. They value the right of the unborn life higher than the woman's right to self-determination. In the joint declaration of both the Protestant and the Catholic churches in Germany, 'God is a friend of life' from 1989 it is stated: 'The woman's right to self-determination cannot be the basis for disposal over the life growing in her'. With this, I reach the second part of the title of this article.

The Right of the 'Other'

The open formulation 'the other' already indicates that we are entering uncertain ground. Is 'this other' already a human being, a child, is it an embryo, a foetus, a life, a subject who has a claim to rights? Which right is meant? Is the embryo a 'she' or a 'he', or an 'it'?

I understand 'the other' to mean the embryo/foetus, and its right as the right to protection and life. The embryo is not able to represent its right and claim itself, but legally and morally it is accorded this right vicariously. Since the middle of the 1980s, artificial fertilisation *in vitro* has been possible in Germany and the 'morning-after pill' is available which prevents the possibility of implantation in the first days after conception. Since 1987, legally from the day of conception onwards the claim of the unborn life to protection is made over against the claim of the woman to self-determination in the case of an unwanted pregnancy.[10] Up until then the time until implantation (ca. the thirteenth day) before which fertilised egg cells would frequently abort spontaneously had not been regulated legally. Besides the 'morning-after pill', other European countries have also admitted the RU 486 pill which terminates the pregnancy in the first days after implantation with antiprogesteron.

Statements from the churches and German legislation both agree since the end of the 1980s that human life and the claim to human dignity begin at the time of fertilisation.

Historically, the unborn was granted the right to life at different stages between conception and birth.[11] Up until the nineteenth century in the Catholic church the date of ensoulment was the measure from when onwards the embryo was regarded as a human being, i.e. for boys from the fortieth day, for girls from the eightieth day onwards. In Europe a punishable prohibition of abortion was only introduced in the context of measures in population policy, the first time after the plague epidemics of the Middle Ages. Until two hundred years ago in Prussia capital punishment was valid, from 1794 a prison sentence from 10 years up to life imprisonment.[12] Up until the middle of the nineteenth century a distinction was made between the 'animate' and the 'inanimate' *conceptus* according to the criterion of perceptible foetal movements. These counted as the date of the beginning of its life; only with quickening the pregnant woman could be sure of her pregnancy. However, up until half of her pregnancy had passed could the woman resort to not having known of her pregnancy. Abortion was condemned as infanticide in Prussia from 1794 onwards if the child was 'alive, formed, in time and complete', i.e. if it was at least '18 inches tall', weighed 'six pounds', had 'hard nails and ears' and a 'reddish-smooth skin' (Jerouschek, 1988, p. 251). The women and those skilled in health matters who procured the abortion were punished.

If the embryo is accorded an unrestricted right to life and full dignity from fertilisation, i.e. from the beginning of genetic individuality onwards, any kind of research, any creation of embryos for research purposes, any germline intervention and any wastage of surplus embryos in IVF is prohibited.[13] Neither is it permitted to distinguish between disabled and not disabled life.

At first glance the dividing line in the title could run between the feminist positions presented and the position of the churches. Both sides disagree in their moral acceptance of a woman decided to terminate her pregnancy. A termination denies the embryo's 'right' to life. A second glance, however, shows the disabled feminists' position to be on the other side, or at least, halfway in between. They promote the reproductive autonomy of women, but understandably deny pregnant women this autonomy when it is directed against a disabled child. Against these women who in principle accept pregnancy, but not in the case of a disabled foetus, they invoke the right of the 'other'.

Even if the question remains open whether the right of the embryo or foetus is exactly equal to the woman's right to self-determination, whether it is given precedence, as in the statements of the churches, or is subordinated to the woman's right to self-determination, considerable contradictions arise with regard to usual medical praxis, genetic research, legal judgments and to theological evaluation. I can only mention them briefly here. They indicate that there is no societal consensus in these questions which can only be solved in a preliminary way.

Does not already prenatal diagnosis itself contradict the dignity and right of unborn life since it serves to segregate disabled life and since with positive test results the medical indication is given which legalises the termination of pregnancy and accepts it morally? I believe that prenatal diagnosis at present in fact primarily serves purposes of selection and that it therefore cannot be regarded as an ethically neutral technique, as it is done in the declaration of the churches of 1997. With their pre-decision for prenatal diagnosis which would lead to the decision for terminating the pregnancy in the case of positive results, pregnant women and couples signalise to society that personally, they cannot imagine a future with a disabled child.[14]

In my opinion, this attitude has to be clearly distinguished from the eugenics of the Nazi period where the abortion of a handicapped child was forced upon parents by the state, where disabled people were sterilised, abused for research purposes and murdered. Today pregnant women and couples decide in favour of one singular, and that is 'their' personal selective termination. They do not discriminate directly against existing disabled persons. But that living disabled people feel discriminated against through this decision, is evident.[15]

With this remark, the social context gets into view. In our society not only families with children are effectively underprivileged, and among them especially single parents, but to a far higher degree families with disabled children, and again especially single parents. Society has to decide in a political process whether it is ready to shape its solidarity systems in such a way that it integrates the disabled, that it recognises handicapped life as an enrichment.

Only then disabled people will be granted the basic rights which the Constitution guarantees to them: 'Nobody may be discriminated against because of a disability'. Are we ready, even if we do not live in direct contact with disabled persons in our families and circles of friends, to integrate them into public life and to look after them financially without being able to expect economically calculable returns from them? Are we ready to share the tasks and pay a recompense for mothers and parents if they are decisive in favour of a disabled child? If we urge women in their pregnancy morally to forego prenatal diagnosis or to accept a disabled child without actively engaging in abolishing the disadvantages which mothers and parents experience through handicapped children and to which these children are exposed, we continue to shift a responsibility which belongs to society as a whole, to the pregnant women and couples.[16]

Finally, the conflict between a woman's right to reproductive self-determination and the right of the other is made acute in one more way. Women whose responsible self-determination manifests itself in their decision to forego prenatal diagnosis and who give birth to a disabled child, are confronted with the claims of others. They are accused of an unsolidaric attitude towards society. Court sentences which demand upkeep for disabled children from doctors who have not informed their patients sufficiently about prenatal diagnosis, contribute to society's lessening willingness to accept disabled life. It is neither the dignity of human life, including the unborn, nor is it self-determination in pregnancy that decide for or against the life of genetically abnormal embryos, but financial cost and efficiency considerations of a society.

Challenges for the Churches and Christian Theology

What can the churches, what can theology contribute to granting women scope for self-determination in a problem pregnancy and to protecting embryos independently from their genetic make-up?

The previous two sections have shown that today, pregnancy is linked to two complexes of problems: the question of termination, and the question of the dignity of unborn life independent of its genetic heritage, in other words, the question of selecting a life that is genetically deemed to be 'superior'. Both issues have an individual and a sociopolitical dimension. On both a heavy historical burden is weighing, and this without molecular genetic methods. Since I consider both embryo protection and respect for the women's right to self-determination as important goods, I shall inquire in this part what the praxis of the churches and theological theory can contribute to granting women scope for self-determination in a crisis pregnancy and to protecting embryos independently from their genetic make-up.

Before I outline a possible contribution of the churches and theology in five points, I have to point out the relativity of these proposals. Women (and a number of men) know that it is extremely questionable to formulate adequate and generalisable statements on pregnancy termination which do justice to the complex situation of women who have decided in favour of termination. Just like the criminal law, rules and moral instructions often are well above the pressing practical and spiritual problems in concrete cases. Correct theological statements lose their credibility and relevance with regard to this kind of conflict. I see these difficulties as one of the reasons for the far-reaching silence of German women theologians on this subject. A number of them have gone through an abortion themselves, many know friends, sisters, mothers or daughters who have had this experience, and the great majority, I suppose, have asked themselves in their imagination whether they themselves would have opted for or against pregnancy termination in an emergency situation.

1 The churches can support pregnant women by offering counselling facilities. In the conflict on whether to undergo prenatal diagnosis and in the decision on pregnancy termination after a positive result counselling agencies of the churches can offer independent and encompassing counselling on the strength of their long experience and the trust that they and other institutions such as Pro Familia have acquired with pregnant women and in society. The existing facility of crisis pregnancy counselling would have to be extended through an offer which focuses on the decision for or against undergoing prenatal diagnosis. For example, women with disabled children, women who underwent prenatal diagnosis and those who did not could share their experiences with pregnant women. They could reflect on their situation in communication with women who know the experience. Before making their decision on a termination the pregnant women should know and weigh up all the alternatives.[17] Before deciding on prenatal diagnosis they should be informed of all the possible consequences.

If the churches extend their presence in this area of public service they can testify to a conviction which would correspond to a theology of Christian freedom, that is that certain issues for decision should be kept free as far as possible from the regulations of the criminal law. They are the areas of decisions of conscience by individuals and personal convictions which cannot be resolved in the pattern of 'more or less'.[18]

2 The churches should continue to engage themselves for sociopolitical measures for children and for disabled people. For several years already the churches have emphasised in their statements that sociopolitical measures and institutional facilities for disabled people and for children have to be improved. Only when women are no longer placed at a disadvantage because they are having and rearing a child alone or in relationship with a partner, and if they are not discriminated against even if they are having a disabled child, can pregnant women in fact come to a decision freely. In actual terms, however, single parents belong to the poorest and most underprivileged members of our society. This constitutes the societal core of the individual problem of pregnancy termination.[19]

3 As in their Declaration for the Week for Life 1997, the churches should promote extensive discussions on the genetic knowledge available in the future, on its application and on the political and moral evaluations of these informations.

The threat which genetic diagnosis and procedures pose to the dignity of embryos needs to be discussed and regulated politically, instead of being focused on the individual expectant mother in the consulting rooms of obstetricians. We need a broad public and political discussion on whether we want to acquire the prenatal genetic knowledge possible, and if so, what kind of use we allow it to have politically. Is there a right to prenatal knowledge? Who is to have it? May it be put at the service of selection? Do we ascribe the same rights to genetically abnormal life as to genetically normal life? Can we permit germline interventions in the face of future generations? In Germany many of these questions up to now are clearly answered in the negative both by legal and church institutions, yet in practice they are decided differently and in fact partly undermined. Moral responsibility cannot be restricted to the realm of national borders. The churches and the feminists demand or defend the right not to know. I consider this important; at the same time it is not sufficient to admit the right not to know to individuals, if they or their children will be disadvantaged by this. Neither is it enough to declare a taboo. Knowledge and technical development cannot be turned back; many women want to use the possibilities of prenatal diagnosis. Politically and legally the point is above all to limit and control the way institutions dispose with this knowledge. Those responsible in firms and insurance companies in the interlinked area of medicine, research and economy, can be obligated to demonstrate that they respect the human dignity of embryos whom they investigate and in whose genetic structure they possibly intervene. On account of the human dignity anchored in the Christian faith the churches have a special responsibility for promoting the public process of debate and for engaging themselves for embryo protection.

4 Theology and the churches are challenged to view the situation of women in pregnancy conflicts in a differentiated way. They should make it clear to the women concerned that God is not only a friend of the weak unborn life, as it is stated in the Declaration of the Churches, but also a friend of the lives of weak living women and weak children; and that God is a forgiving God who accepts people who have become guilty. Theological theory itself is challenged to support the lives of women in a pregnancy conflict.

This happens first by taking them seriously and respecting them as subjects of the conflict. The decision on what women can afford to ask themselves, their children and families to bear, cannot and should not be taken away from them in a liberal democratic society and in a church that invokes Christian freedom. This is why it is crucial that the woman in conflict can make her irreplaceable decision of conscience as freely as possible. When a woman decides in favour of terminating the pregnancy or against a disabled child, it should not be presupposed in advance that this is done light-heartedly. Instead, women act in greatest desperation under enormous time pressure and in responsibility towards the growing life, towards other children, towards themselves. They are caught in a situation from which there is not good way out, and opt under time pressure and in justified anxiety for the alternative which at that time appears a little less bad than the other. Their anguish is increased through criminal law and moral precepts from the churches.

To recognise this theologically, does not imply in my opinion to avert one's eyes from the fact that to terminate a pregnancy means to kill life. In addition, the decision against a disabled child is eugenic selection which is contrary to God's love to the disadvantaged and to God's love of life.

To have denied possibilities of life to their unborn child and have denied themselves and the child the chance of a shared life, often burdens women for a long time after the termination. Many despair, are filled with deep grief and feel guilty.[20]

It would be important if the churches also realised that it is not only pregnant women individually who are responsible for the difficulties surrounding a crisis pregnancy which does not offer any good way out. In a society hostile to children and even more hostile to disabled people many people are participating in the context of guilt: beginning with the child's father, to the landlords who do not accept families and even less families with handicapped children, up to a legislation that enables companies in Germany to buy themselves free from the prescribed quota of 6 per cent of workplaces for disabled people. It is the pregnant women who take over their particular suffering from this context of guilt in a vicarious way for many others, by undergoing the termination and the severe intervention into their bodies, or by having to rebuild their life projects as future mothers of disabled children, sacrifice their professional lives and being available for their children around the clock and far beyond their childhood. If they decide in favour of pregnancy termination, they accept to be injured profoundly both physically and spiritually because they consider the suffering that would ensue if they did not do it to be even greater.[21] The responsibility for themselves and others, out of which they make this decision deserves respect and acceptance.

For Christian theology, the knowledge is constitutive that human beings are in need of forgiveness. It can make life possible again and heal wounds. Theology should make it clear that women who suffer after pregnancy termination from self-reproaches and guilt attributions from others are the ones to whom this forgiveness applies. It can help them to forgive themselves and to grant themselves the chance of a new beginning.

Apart from systematic-theological reflections, liturgical forms should be found which can give ritual expression to the sorrow for the life killed, to the expression of despair, the confession of guilt, the divine promise to renew life and the announcing of forgiveness. Special services in the congregation could offer the woman or the couple space before God and provide rituals to part with the unborn and express their grief. The communal liturgy could show that not only the pregnant woman or the couple are affected by the termination. Only if these liturgical forms are not focused on the individual woman personally but take the structural and personal failure of all those directly and indirectly responsible into view, can they do justice to the real problematic.

5 It should be reflected theologically whether it would be helpful to re-emphasise the bodily dimension of pregnancy and birth in the theological definition of what it is to be human.

The right of the unborn life over against the pregnant woman and the protection of embryos against research, destruction and germline interventions are closely linked in Church and legal argumentations. What joins both concerns is the definition of the embryo from fertilisation onwards as an individual human being who participates fully in human dignity. This dignity which is granted by God from birth onwards until death and guaranteed in human declarations of self-obligation since the end of the 1980s in German jurisdiction is extended to the embryo from the time of fertilisation onwards. Nevertheless, the Protestant Church and also legislation show a certain amount of understanding for the termination of a pregnancy while on the other hand, they give unrestricted support for embryo protection in the context of research. This position which could seem inconsistent might indicate that the 'maximal definition' of the embryo's dignity may after all have problems in the practical sphere.

A theology that highlights the bodily constitution of the human person as a crucial dimension, as the Declaration of the churches 1997 does, should not ignore the physical experience of pregnancy: the bodily fact that the pregnant woman and the growing child form a dyadic union. Pregnancy is a process the beginning of which is marked by a woman who is not yet sure whether she is pregnant; at the end of it there is a mother and her daughter or her son. In this process the first foetal movements and the birth are of special significance. These 'steps' or 'thresholds' could be recognised theologically. That human life begins before birth, that it enjoys dignity and the right to protection does not have to be put into question. This bodily perspective does not have to replace the other, genetic-biological view. It explains why it is the woman whose body carries and nurtures the unborn who has the prime responsibility in a problem pregnancy before all other agencies. It supports the intention of the churches and of jurisdiction not to enforce the right of the unborn against the will of the pregnant woman. It allows to differentiate between a pregnancy termination in the first months and the killing of a human person.

The precarious situation of the embryo *in vitro* would become especially clear from this perspective. Since the embryo has been taken out of the protective space of the mother's body, it needs greater attention and vicarious public protection, because it is the being that can be genetically manipulated.

A body-bound perspective on pregnancy juxtaposes the two rights mentioned in the title. Today, the right of the 'other' is not so much threatened by women's right to self-determined pregnancies but through human genetic research and application which does not have sufficient democratic control. This threat not only affects 'the other', but ourselves and our children.[22]

Notes

1 I am treating the rights of the woman and of 'the other' in the following reflections not in a legal sense but from the perspective of a Protestant German theologian. I am therefore speaking of rights in the sense of moral claims which demand recognition.

2 In amniocentesis the average risk of chromosomal abnormalities in women around the age of 35 years is at 1:385, the risk of miscarriage is at 1:100; in chorion villus sampling the rate of miscarriage according to Schindele ranges from 3.5–7.2 per cent and the mistake rate at 1 per cent. There are no figures given with regard to neural tube defects. Figures according to Schindele, 1995, pp. 162–79.

3 The wording is:

> The termination of a pregnancy through a physician with the consent of the pregnant woman is not against the law if the pregnancy termination is indicated according to a medical assessment which includes the present and future life circumstances of the pregnant woman, if it is done in order to avert danger for the life or danger of severe harm for the physical and psychological state of health of the pregnant woman and this danger cannot be averted in any other way which she could reasonably be expected to support (*zumutbar*).

Quoted after Kirchenamt der Evangelischen Kirche in Deutschland/Sekretariat der Deutschen Bischofskonferenz, 1997, p. 6.

4 Until 1995 in Germany in these cases an embryopathic indication was issued which referred to the embryo's state of health. This indication which was openly eugenically motivated has been legally abolished, partly because of the criticism of the associations of disabled persons; yet it actually has only been passed on under the title of medical indication. Many critics see the main problem in the new regulation of Paragraph 218 in the fact that there is no time limit for the medical indication, i.e. that it can be issued until shortly before birth (cf. Huber, 1996, p. 356). Dietmar Mieth (1997, p. 249) problematises the legalisation of pregnancy termination with medical indication.

5 The suction method is more gentle for the pregnant women, yet it dismembers the foetal tissue.

6 This is what Dr Vieth told me in a telephone conversation on 23 February 1998.

7 According to an article in *Spiegel* magazine of December 1997 recently there have been successful prenatal therapies of accelerated heart beat and high blood pressure, while experimental therapies of spina bifida failed in the US (*Spiegel*, 52/1997, p. 25). (Translator's note: In November 1998, successful prenatal therapies also of spina bifida were reported from the US.) Jörg Schmidtke states that the regular intake of folic acid during the first 12 weeks of pregnancy reduces the risk of neural tube disorders and adrenogenital syndrome for genetically affected couples from 4 to 1 per cent (Schmidtke, 1997, pp. 276–9). Further illnesses which can be therapied are heart insufficiency and urinary tract obstructions.

8 Apart from her critique of any societal or state interference into the issue of pregnancy her goal is to initiate among women 'a collective process of reflection about bearing and having children'. It is not to be limited either to contraception and abortion nor to gentle births and the new motherliness but to include 'the whole range of possible dangers and complications as well as the entire weight of responsibility of women's decision over the life and death of an embryo/foetus/child' (Kontos, 1986, p. 42).

9 Kuhlmann interprets the court decision differently. He expects from physicians comprehensive counselling if women desire it so that an independent decision becomes possible (Kuhlmann, 1996, p. 207).

10 According to the Declaration of the Churches also the embryo in the mother's womb from conception onwards is 'a different human being of its own' to whom human dignity is attributed (Kirchenamt der EKD/Sekretariat der Deutschen Bischofskonferenz, 1989, p. 44).

11 Here, different biological-medical argumentations are given. Besides the time of conception, the seventh day as the end of the possibility of the formation of twins could be counted as the beginning of human life, the beginning of brain activity at the end of the sixth week, the date of the first foetal movements experienced by the pregnant woman between weeks 14 and 20, week 24, when chances of survival in the case of birth exist, or birth itself (Körner, 1992, p.122).

12 In order to erase the knowledge of contraception as well as of methods of abortion of the 'wise women', these were demonised and persecuted as witches up into the eighteenth century. The history of the prohibition of abortion from antiquity until the nineteenth century has been documented and commented on by Günter Jerouschek (1988).

13 Besides cloning, genetic technology has recently realised the production of animal organisms without heads as pure organ stores. In principle, this can also be applied to human beings. With regard to the great lack of organs for transplants the provision of human organ stores is being discussed internationally. In Germany and also in the Convention on Human Rights and Biomedicine the taboo with regard to these forms of human genetics is upheld.

14 However, a number of women resp. couples whose developing child has been diagnosed with a less severe disability, decide in favour of continuing the pregnancy. Kuhlmann cites the Turner-, the Klinefelter- and the Triple-X-syndrome as examples of such less severe disabilities (Kuhlmann, 1996, p. 151).

An expertise of the Parliamentary Bureau for Assessing the Consequences of Technology (Büro für Technikfolgen-Abschätzung) states:

> We did not find any suggestion in our counselling clients that women only want healthy children and only accept them if they are healthy, nor that women are ready to terminate their pregnancy on account of any positive result, no matter how insignificant ... After counselling, women do not terminate for trivial findings; also pregnancies where sex chromosome abnormalities have been found are only terminated if there are positive sonographic results (Nippert and Horst, 1994, p. 100, quoted after Kuhlmann, 1996, p. 231f).

15 'The signal "We don't want someone like you!" – and this signal doubtlessly arises from the practice of selective abortion – can be experienced by many [disabled people] as a personal humiliation' (Kuhlmann 1996, p. 165). The interest in selection is even more unrefutable with pre-implantation diagnosis. If a genetic abnormality is found with an embryo conceived *in vitro*, it is not implanted into the uterus.

16 Evidently the decision remains an individual and subjective one. My concern here is to point out the social conditions and backgrounds of the individual decisions. Subjective reasons which are projected onto the otherness and abnormality of disabled people that result in the termination of the pregnancy and a hostile attitude towards disabled persons are investigated by Susanne Ehrlich (1993).

17 They should also take into account the possibility of adoption which, however, in the case of disabled children is only offered in very rare cases.

18 To me, it appears as a matter of course that this counselling is in the service of life. At the same time I think that its result should be open in the sense that the pregnant woman decides herself after counselling if she is better able to support a small measure of disability than choose the alternative.

19 It is still the case that the proportion of families with children constitutes the greatest part of recipients of social welfare; of these, single mothers and their children form the highest number. The cuts in social affairs which neither spared institutions for disabled people nor social welfare, however, give reason for pessimism.

20 Several research reports show that by far not all of them feel guilty (e.g. Knopf et al., 1995).

21 This is to be distinguished from the readiness to take over guilt which Dietrich Bonhoeffer demands in situations of conflict; he refers above all to the problem of the murder of a tyrant (Bonhoeffer, 1992, pp. 275–83). Also pregnancy termination is a situation in which every decision incurs guilt. Yet women who decide in favour of terminating a pregnancy do not heroically take over the responsibility for all to kill, in order to save all. They take over this woundedness because they regard the suffering which the decision to give birth to the child would cause as greater for their personal environment, their family and their own lives than a termination.

22 I want to thank Hildburg Wegener, Dieter Weber and Ilona Nord for important bibliographical references and Dr med. Cosima Vieth for two extensive advisory and critical conversations.

References

Bonhoeffer, D. (1992), *Ethik*, München.

Degener, T. and Köpsell, S. (1992), *'Hauptsache, es ist gesund'?, Weibliche Selbstbestimmung unter humangenetischer Kontrolle*, Hamburg.

Duden, B. (1991), *Der Frauenleib als öffentlicher Ort. Vom Mißbrauch des Begriffs Leben*, Hamburg/Zürich.

Ehrlich, S. (1993), *Denkverbot als Lebensschutz? Pränatale Diagnostik, fötale Schädigung und Schwangerschaftsabbruch*, Opladen.

Hille, H. (1995), 'Historisierender Feminismus als Regression. Zum Ressentiment gegen die Moderne bei Barbara Duden', *Zeitschrift für Sexualforschung*, 8, pp. 249–58.

Huber, W. (1996), *Gerechtigkeit und Recht. Grundlinien christlicher Rechtsethik*, Gütersloh.

Jerouschek, G. (1988), *Lebensschutz und Lebensbeginn. Kulturgeschichte des Abtreibungsverbots*, Stuttgart.

Kirchenamt der EKD and Sekretariat der Deutschen Bischofskonferenz (eds) (1989), *Gott ist ein Freund des Lebens. Herausforderungen und Aufgaben beim Schutz des Lebens*, Gütersloh.

Kirchenamt der Evangelischen Kirche in Deutschland and Sekretariat der Deutschen Bischofskonferenz (eds) (1997), *Wieviel Wissen tut uns gut? Chancen und Risiken der voraussagenden Medizin*, Gemeinsame Texte 11, Hannover/Bonn.

Knopf, M., Mayer, E. and Meyer, E. (1995), *Traurig und befreit zugleich. Psychische Folgen des Schwangerschaftsabbruchs*, Reinbek bei Hamburg.

Kontos, S. (1986), 'Wider die Dämonisierung medizinischer Technik' in Die Grünen (ed.), *Frauen gegen Gen- und Reproduktionstechnik*, Köln.

Körner, U. (1992), 'Wann beginnt das menschliche Leben?', *Ethik in der Medizin*, 4, pp. 120–34.

Kuhlmann, A. (1996), *Abtreibung und Selbstbestimmung. Die Intervention der Medizin*, Frankfurt am Main.

Mieth, D. (1997), 'The Market and the Inviolability of Human Dignity in the Perspective of Biomedical Technology' in D. Mieth and M. Vidal (eds), *Outside the Market no Salvation*, Concilium 1997/2, SCM Press, London.

Nippert, I. and Horst, J. (1994), *Die Anwendungsproblematik der pränatalen Diagnose aus der Sicht von Beratenen und Beratern. Gutachten im Auftrag des Büros für Technikfolgen-Abschätzung beim Deutschen Bundestag*, TAB-Hintergrundpapier No. 2, Bonn.

Schindele, E. (1995), *Schwangerschaft. Zwischen guter Hoffnung und medizinischem Risiko*, Hamburg,

Scholz, C. (1995), 'Biographie und molekulargenetische Diagnostik' in E. Beck-Gernsheim (ed.), *Welche Gesundheit wollen wir?*, Frankfurt am Main.

Schmidtke, J. (1997), *Vererbung und Ererbtes. Ein humangenetischer Ratgeber*, Reinbek bei Hamburg.

Translated by Maureen Junker-Kenny from the published version of the paper in *Zeitschrift für Evangelische Ethik* (1998), 42, pp. 266–81.

7 Ethical Reflections of Genetic Counselling in Prenatal Diagnosis

HILLE HAKER

The Current Approach to Prenatal Genetic Counselling

Within the last 20 years, prenatal diagnosis has been established in many European countries as a tool of monitoring pregnancies. During the 1970s and 1980s, it was considered to be helpful for couples who otherwise would refrain from reproduction fearing the birth of a disabled child because of a specific family history. Furthermore, it was held that prenatal diagnosis would help avoiding abortions for the same reason. The addressees of prenatal diagnosis were therefore women or couples who ran a high risk to have children with genetic disorders that could be diagnosed during pregnancies. The identifications of these high risk couples was made possible either by family studies of monogenetic diseases or by specific genetic testing of the couples after the birth of a disabled child.

This group of high risk clients was soon expanded to women from 35 years on. The family-indication and the age-indication again broadened, when prenatal diagnosis became more accepted and non-invasive tests were introduced: above all the so-called triple-test made it possible to record those women whose statistical risk was higher than the age-factor would indicate. Although the Triple-test has been criticised by many human geneticists, it was nevertheless established. Together with ultrasound diagnosis and the alpha-fetoprotein test it can be considered as an indication for invasive prenatal diagnosis.[1]

Since the 1990s, the group of addressees of prenatal diagnoses has been expanded again to those women who articulate their fear to give birth to a child with genetic disorders without having family-indicated or age-indicated risks. Today, we also confront the question of genetic screening, as it is recommended, for example, for the triple test in several countries, or for cystic fibrosis (recommended in the USA since 1997, after positive evaluation of a pilot study) (Witt, 1996).[2] The debate on prenatal diagnosis as an instrument for sex selection, as it is seen in some states, e.g. India or China, indicates that another broadening of diagnostic measures is taking place: the extension to non-disease-related diagnoses.

Along with this progression of the *medical* practice, genetic counselling was developed as an accompanying offer to help women or couples make their decisions about diagnostic measures. The beginning of genetic counselling, in the first half of the twentieth century, can be summarised in the terms of *prevention* and *directivity*. The obvious eugenic concept of preventing offspring of 'genetically inferior quality' led to directive counselling, advising couples or parents to voluntarily refrain from having children. Furthermore, compulsory means, as sterilisation, were considered as medical treatment to fulfil the eugenic preventive goal. From 1945 to 1970, this approach was continued, although the compulsory state intervention having been practised by the German Nazi regime had become problematic. Nevertheless, the Nazi practice was rather considered as a criminal excess of a valid paradigm rather than questioning the paradigm of prevention itself (Haker, 1993). This changed in the 1970s and 1980s, and with the new development of genetic intervention, new concepts of counselling were being developed, resulting in the principle of nondirectivity. However, up to the present there is a tension between the preventive goal of genetic diagnosis and the principle of nondirectivity which is based on the freedom in reproduction or 'reproductive choice' as a private matter of couples. There are many attempts to replace the goal of prevention and substitute it by the goal of 'help in family planning', but since the latter, too, is ultimately resulting in prevention, it has neither been successful in theory nor in practice.

The concept of genetic counselling, as it is presented since the 1970s, is based on several maxims that can be summed up in the terms of *information*, *nondirectivity* and *voluntariness* (Bartels, 1993; Baitsch, 1986). It is aimed at a decision-making process in which clients should have access to the information that is needed to make an *informed consent* to the diagnoses, and in which they should not be directed in whatever way the counsellors might wish them to decide, so that their consent could truly be called a *free consent*. The ethical principle that is referred to by this concept is the respect of autonomy of the client or patient, although the three other well-established principles of medical ethics: beneficence, non-maleficence and justice, play a secondary role in the background of the counselling concept.

When the ethical debate about prenatal diagnosis began to intensify during the 1980s, it was the concept of genetic counselling that was held up against criticism. Genetic counselling could now be seen as the communicative part of medical help for families, but as well as professional fig leaf in ethical discussions. Today, the concept of prenatal genetic counselling is widely accepted not only as an offer, but also as a condition for genetic diagnosis, at least in its invasive forms. It is supposed to exclude arbitrary parental decisions about the termination of pregnancies, and to take seriously the women or couples as moral agents who are supported in making autonomous decisions.

Seen in the historical context of human genetics, this approach can certainly be regarded as progress, since it tries to take seriously at the same time the concerns of the clients and the ethical problems that are involved in the decisions about the life of a foetus. However, it must be asked if the ethical problems are faced *adequately* in this concept, and if the dominance of the principle of autonomy is ethically sound.

Autonomy of the Women/Couple and the Benefit for the Foetus

The principle of autonomy plays a central role in current ethical discussions about prenatal diagnosis, since diagnostic measures are seen in the context of a couple's reproductive choice (Schöne-Seifert and Krüger, 1993). Along with the technical development and the establishing practice the option of choice was expanded, too. Taking the couples' 'free choice' in reproductive matters as starting point of the moral reflection, the limitation of prenatal diagnosis by setting up medical indications appeared to require strong arguments. Besides practical reasons for limitation of access, e.g. availability and financial costs, human geneticists and ethicists have argued that the concern for the foetus has to be considered likewise. Yet, geneticists struggle with the problem of unlimited access in another way: as the number of clients is steadily growing, the establishment of genetic counselling centres or professionals who are educated in genetic counselling has not been growing respectively. Geneticists fear the undermining of their concept that prenatal diagnosis has to be combined with genetic counselling (Schroeder-Kurth, 1993; Wolff and Jung, 1994).

However, ethicists who strongly support the establishment of prenatal genetic diagnosis argue that a limitation of access would easily get into conflict with the principle of autonomy, namely the more specific 'right to know'. They indicate that couples or parents are the ones who can decide not only in their own best interest, but also in the foetus's best interest, and that no other person or institution should put pressure on them in one direction or the other

(Birnbacher, 1993). The supporters of the principle of autonomy therefore implicitly presuppose that the foetus's life is in the hands of the couple or parents, and that it is not a good that has to be protected independently by institutions.

In my own view, the concentration on the principle of autonomy, even if it is balanced to questions of a just distribution of financial resources, conceals the underlying ethical problems rather than faces them. For what is at stake in prenatal diagnosis – as long as it is not followed by a prenatal therapy which has its own problems – is the decision about the life of a human being that is not yet born but shares all essential physical features of a newborn child. This decision is not only to be made in the tragic cases of diagnostically positive results, but it is present from the very beginning of the decision-process to undergo prenatal diagnosis. Furthermore, it is not only the decision of the clients that needs ethical reflection, but also medical practice.

For with the development of pregnancy monitoring the foetus has become a *patient* whose physical and genetic constitution is diagnosed. Along with this development, the medical profession becomes responsible not only for the pregnant woman, but for this patient, too. According to the principles of medical ethics, especially the principle of beneficence and non-maleficence, no medical action may be tolerated that has negative consequences which cannot be made up by positive outcomes to the same subject. In short, this is a direct consequence of the concept of human dignity, stating that no human being may be entirely instrumentalised for the good of another human being. The application of these principles of beneficence and non-maleficence that are reflected and specified in many directives in the specific fields of medicine, certainly is complicated in the case of prenatal diagnosis, but still it seems obvious that a departure from them has to be argued very carefully.

In prenatal genetic testing, diagnostic techniques are offered that until today include serious risks for the foetus – above all chorionic villus sampling and amniocentesis, which are the common *invasive* techniques of prenatal diagnosis.[3] To the present, these risks cannot be balanced against possible therapies. As long as prenatal diagnosis is based on invasive techniques and as long as there are no (or almost no) therapeutic measures to go with them, we face the problem that I would like to call ethical efficiency. By this I mean that the good intended with a certain technique has to be clearly defined. Only if the possible ethical side-effects are justified with respect to the good, an action can be called ethically efficient and, along with it, justified.

In the present discussion the intended good of prenatal diagnosis is considered to be the expansion of the couples' or parents' autonomy with regard to their family situation and their future child or children. Prenatal diagnosis is seen as an alternative to a family life without children or a life with the burden of rearing a severely handicapped child, or an abortion on spec. Furthermore, some ethicists argue that avoiding a life of severe suffering would benefit the future child itself (Harris, 1993). Thus, the intended good of prenatal diagnosis is defined both by the presumed interests of the foetus and the articulated interests of the woman or couple. If there is a moral right of the foetus, namely the right to life and the right to be protected against harms – which some ethicists doubt – it is argued that this right can still be overruled either by the foetus' own interest to be protected against an existence that would mainly consist of suffering, or by the parents' right to their life, consisting of a quality that only they themselves can determine. However, the specific normative character of the foetus's right is hardly ever worked out.

The argument that prenatal diagnosis was at least partly submitted for the benefit of the foetus itself presupposes that there are diseases so serious that prolonging of pregnancy would do the foetus no good, especially in the case of an early death after birth. The difficulty of this argument is obvious: it makes it necessary to define the borderline cases and the quality of the suffering in objective terms. Besides this it can be questioned that the present practice of prenatal diagnosis is restricted to these rare cases.

Human geneticists and the medical profession up to the present have hesitated to follow this path. Instead they approach the problem from a different perspective that emphasises the suffering of the mother and/or parents of the child. The departure from the common medical principles then is seen in the specific constellation of pregnancy and the relationship between mother/parents and child. The foetus's dependence on the pregnant woman reversely reflects her responsibility for the foetus, a responsibility that is prolonged in the parental responsibility for the child. Faced with the risk or, after diagnosis, with the certainty of a genetic disorder, the woman's and parents' life is as much at stake as the foetus's life. It is argued that in the borderline cases where termination of pregnancy appears to be the only possible solution for the woman or couple, their decision has to be accepted, however supported by genetic counselling ideally offered before and after diagnosis has taken place.

However, genetic counselling faces several problems: first, *counsellors have to define their specific role*: counsellors could compare themselves to psychotherapists whose counselling is 'nondirective' in that sense that they help their clients to clarify their life-goals and life-decisions, but leave the decisions to their clients. This is the background approach of Rogers that was implemented in the concept of nondirective genetic counselling during the 1970s.[4] But are questions of prenatal diagnosis comparable to other life-decisions that are more or less centred on a person's concept of his or her life, ethically described in the term of 'questions of good life'? Or does this imply the presupposition that no normative and moral questions are involved in prenatal diagnosis? Thus, it is held that counsellors should rather compare themselves with physicians who have a specific competence in a course of therapeutic actions that their patients might not be able to judge adequately. A certain kind of directivity cannot be excluded in this approach, even though it is embedded in the 'free and informed consent' of the patient. Yet there is another approach to define the role of the counsellor, which takes seriously the specific relation of the physician to the two patients he or she must recognise. If the foetus is considered a patient, then the paediatric approach of advocative decision-making appears to be the adequate model. In this approach, the counsellor will define himself as mediator between the two subjects, the mother or couple and the foetus. To follow the paediatric approach implies respect for the foetus regardless of the interests of the mother, as long as her life is not endangered by the foetus itself. In the German legislation, for example, termination of pregnancy is not against the law in the case of a so-called medical indication. Since 1995, however, this specific indication that protects the health and life of the pregnant woman, also applies to a detected disease of a foetus. The law states that only if the diagnosed disease results in a serious physical or psychological impairment of the mother's health, termination is legally allowed. Although the interpretation will cause a lot of trouble, since it is the future health of the mother that is at stake, the intention of the law is to give the foetus as much protection as possible. The mere identification of a genetic disease cannot be the reason for a termination of pregnancy. Following this approach, the interpretation of the mother's health cannot be exclusively made by herself but will be the result of a communication process between her and the physician and/or counsellor.

Thus, the self-understanding and definition of the role of the counsellor is decisive for the whole setting of the counselling.

Another difficulty is the *role of the client*. If a pregnant woman is supposed to be the subject of decision-making, it is not at all clear that she knows how to decide. She may have a feeling of a moral conflict – or not. The principle of nondirectivity is contradictory if the setting suggests that prenatal diagnosis is a private choice, but denies certain requests, as for example termination for non-medical reasons. All Western countries deny such a right, arguing in this case that the request is morally not legitimate. All restrictions either of access or of specific diagnoses are directive in the sense that counsellors or medical institutions impose a certain understanding of respect for the foetus on the woman. But if this is so in some cases, it must be argued why it is not so in other cases.

In the following, I will try to argue for a concept of genetic counselling which integrates ethical reflections without being directive in the sense of restricting the right to autonomy. Before I can do that, however, I will argue for a concept of personhood which allows us to understand and justify the moral protection of human embryos or foetuses. According to this respect it is necessary to define the role of the counsellor as mediator of woman and foetus. Of course, this role of a mediator appears strange in a more traditional understanding of an exclusive relationship of mother and child during pregnancy, and it can also be criticised from a feminist perspective. However, since prenatal diagnosis is in fact an interference in pregnancy, it should be made explicit and reflected upon in a normative sense. On the other hand, a genuine interest of the pregnant woman to protect her child can be assumed in most cases so that there will not be a conflict between her and the physician. Since the motives for prenatal diagnosis are complex, and prenatal diagnosis is at least partly requested to promote the intention to act responsibly, the specific kind of responsibility must be discussed. In the nondirective approach of genetic counselling this ethical dimension is not at all clear. With the internal goal of giving 'neutral' information before and/or after the diagnosis so that the woman can decide what to do, the complexity of the communicative dimension of genetic counselling is not met adequately.

Dignity, Personhood and the Moral Status of the Foetus

Thus, if recourse to the principle of autonomy spelt as respect of the reproductive right is not sufficient for a full understanding of the normative nature of the decisions to be made with respect to prenatal diagnosis, we are taken back to the question of the moral status of the foetus that has to be clarified in order to describe the ethical criteria for justified medical actions on him or her during pregnancy. For it is this question that is decisive for the approach of genetic counselling.

In comparison to the question of the moral status of human *embryos*, it seems to me, that in the case of human foetuses we have better and more valid criteria that indeed allow an answer on better grounds. First, the analysis of the concept of the person discloses well-known aspects of ethical relevance that nevertheless have rarely been applied to the context of prenatal diagnosis; second, the biological stage of development of a foetus in the second term of pregnancy can be considered as a specific basis for the determination of the *foetus's* moral status.

A Phenomenological Approach to Personhood

In the modern tradition of moral philosophy, the concept of personhood interpreted in the term of self-consciousness and the capacity to be an agent forms the basis for the specific dignity of human beings which is explicated in the normative terms of self-determination and responsibility for the actions a person undertakes (Locke), or moral autonomy and non-instrumentalisation (Kant). Dignity is thereby connected with the normative understanding of personhood presupposing that the subject of dignity is a moral agent (Gewirth). Moral rights, as well as moral obligations, depend on this capacity to be a moral agent. Thereby, the concepts of personhood, moral rights and dignity refer to and are defined by each other. However, the understanding of personhood in terms of autonomy, self-determination or agency appears to be too narrow, if regarded from a phenomenological perspective. For the phenomenological basis of personhood brings to the surface some other dimensions that have to be interpreted in ethical terms, too.

First, there is the relation of personhood and time: persons (and by this, moral agents) do not exist as timeless entities, but they exist in their different developmental stages. In different contexts, e.g. age-related dementia or brain damages after accidents, persons may lose their capacities to be moral agents, but nevertheless they are perceived as persons. Furthermore, persons perceive themselves as having been persons all along and to be persons all along in their future, even though they (can) know that they have not always been and possibly will not always be moral agents. It is important to see that although continuity is *a priori* presupposed as an integral part of personal identity, it still has to be created *a fortiori*; this is done in persons' life-stories that present and reflect their identity in the narrative form of succession in time (Ricoeur, 1990; Haker, 1998). Certainly, a narrative of one's own life is possible only if the capacities of self-reflection, memory and self-determination has been developed, but these capacities put us also in a position to reflect about the limits of objectivising our identity. In perceiving another person, we are bound to presume that he or she, too, *is* not only a person but has a history of past and future that cannot be fully explicated or grasped in rational terms. Therefore, there is an unexceedable gap in *being* and *understanding* ourselves and others in their identity. Put in ethical terms we can say that the respect of the other must include not only the respect of his/her present autonomy, but also the recognition of the individual life-history, and this with respect to someone's possible future and past.

The second dimension is the corporeal constitution of personhood. Similar to the temporal constitution of personhood, corporeality can be reflected upon, but we cannot entirely objectivise ourselves as bodily beings. We *are* bodies before we *know* that we are and what we are. The ethical relevance of corporeality can be interpreted in several ways: being persons we are still beings who can be harmed and who can do harm to others in their bodily constitution. To perceive another person means to perceive him or her as a potentially vulnerable being. It is not by accident that in one of the most important phenomenological interpretations of morality, in the ethics of E. Lévinas, it is the perception of the face of the other that constitutes the moral claim to respect the other as another. Second, in being autonomous persons or moral agents, we have to reflect on the corporeal constitution of ourselves with respect to the goods we strive to. Accordingly, in acting on others we have to consider their needs and wishes as well and as much as ours. The similarity we recognise in perceiving another *human* being makes it necessary for us to presume that this other has similar needs and similar wishes to ours. On this basic intuition the Golden Rule is based. It is correct as a first step of moral reflection on the other, and it is correct for those cases in which we cannot correct our perception and assumptions of the other by asking him or her what they really need and what they actually wish for themselves.[5]

The third dimension of personhood is its social roots: personhood is not constituted by a moral agent him- or herself alone, but is at least partly the result of social interaction. This means that every person as a moral agent is or at least has been partly dependent on others to become a person, not only in his or her genesis, i.e. socialisation, but as well in his or her social status. If we take this seriously, the intellectual capacity is not the only relevant factor to become or remain a person, but the intersubjective recognition of a person as a person is constitutive, too (Habermas, 1988; Honneth, 1993). In being moral agents we have to reflect on this responsibility and the power we have on others. The history of social interaction might easily be interpreted as a history of disrespect of the other in terms of inequality that was socially legitimised. The moral claim urges us to reflect on these social constructions and the potentially excluding power of moral concepts, too. Recognising that personhood is at least partly ascribed by others, then, makes it necessary to determine the criteria of this ascription. Concentration on moral agency or self-determination may easily appear as an ideology, if taken as the only criterion to ascribe moral rights. For one important conclusion of the phenomenological interpretation is that personhood begins earlier than moral agency, and that it is not ending with the capacity to act autonomously either. Second, the protection of vulnerability which is a direct effect of the corporeal constitution of personhood must be determined as a right independent of the capacity to claim it. Third, the constitution of personhood in social interaction not only shows the normative necessity of a concept of social recognition, but also reveals the historical fragility of social constructions with their excluding and discriminating potential.

The freedom which moral agents have and rightly claim, includes morality itself as the capacity to take over responsibilities. Dignity, therefore, is rightly considered as a term to mark this status of persons as moral agents. However, the rights that go along with this dignity are broader, as all conditions to be a person have to be considered respectively. In this way, the moral rights' concept can refer to *human* dignity presupposing that moral rights are constitutive to become a person in the only experienced sense of *human* freedom and self-determination, namely within the constraints of time, corporeality and social interaction.

No artificial division of personhood and humanity is necessary, if we consider the dimensions of continuity, corporeality and social recognition as integral parts of personhood. Likewise, the normative grounds of the concept of human dignity is not humanity itself, as it is regarded in the critique of speciesism, but the constitutive elements of personhood itself. In consequence, dignity is ascribed not only to those beings who are moral agents but also to those beings who share at least one of the constitutive elements of personhood. To clarify and differentiate the specific normative respects of personhood is a direct consequence of the concept analysis of personhood. The advantage of this approach is that the interrelation of the physical conditions of personhood and self-consciousness and agency that go along with it can be interpreted not only descriptively but also normatively which is extremely important with respect to the ascription of moral rights.

The Moral Status of the Foetus: The General Argument

With respect to the general concept of personhood there certainly are cases that are difficult to interpret. One of these cases is the determination of the moral status and explication of the rights of the foetus. The general argument, following the rich concept of personhood, interprets the factual status of a foetus as a necessary phase in a person's development. From the perspective of the moral agent's self-reflection it can be held that no person has become a person without having been a foetus, and saying this we interpret the stage of the foetus as part of our continuity concept of personal identity. Furthermore, the foetus, we can say, is a corporeal being which he or she shares with all persons. Finally, not yet being a moral subject, a foetus is overall dependent on another person who is a moral subject: dependent on his/her mother to be protected against harms, dependent on a society to be protected and supported in his moral status. In this respect, moral agents can consistently hold that a foetus is not yet to be respected as a person *in the sense* of being an autonomous moral subject. However, they cannot hold that a foetus does not have the right to be protected in his/her corporeal vulnerability that we see in accordance with our own vulnerability. To deny a foetus that moral right would be a practical misunderstanding of the meaning of personhood. But it would also be a conceptual misunderstanding presupposing that moral rights are ascribed only with respect to moral agency or self-consciousness, and not with respect to the other constitutive elements of personhood.

The Moral Status of the Foetus: The Specific Argument

This view that is applicable to all stages of human life (including the stage of the embryo) is supported by the more specific argument, taking into consideration the biological and physical status of a foetus that is submitted to prenatal diagnosis. To me, three factors of development seem to be relevant here. In the second term of pregnancy, a foetus's organic differentiation is completed. Through sonography, this can be visualised and thereby the foetus is perceived as a being that shares the features of humans. This underlines the proximity to new-borns we know to be helpless and needy beings. Second, a foetus in about the twentieth to twenty-fourth week of pregnancy can be presumed to be capable of feeling physical harm. In the tradition of moral philosophy, this is a decisive criterion to protect a being against harm. Third, at least at present, late abortions take place in a period when principally the survival *ex vivo* would be possible, if intensive care were offered to the newborn. Premature new-borns from about the twenty-second week on are in fact submitted to intensive care, with a success rate of survival that is improving rapidly. The so-called late abortions are actually much more comparable with deliveries of premature babies than with non-medical indicated abortions up to the twelfth week. If the determination of the moral status of foetuses has to be based on analogy, these three factors together show that foetuses can more easily be compared to newborn babies than to embryos or even foetuses in the first term of pregnancy. With newborns they share all relevant developmental factors, including the need to be protected in their previous development. If there is a need to argue the social ascription of moral rights on behalf of the foetus, these comparable factors are a good basis to do so. In consequence, the argument to balance these rights to the rights of the mother/ parents at all has to be much stronger than is suggested in ethical discussions where the right of the foetus is seen more in the context of embryo's rights than children's rights, and prenatal diagnosis seen more in the context of reproductive self-determination of women or couples than in the context of advocatory decisions in neonatology or paediatrics.

Conflict of Rights: The Parents and the Foetus

I have stated above that in the process of modern medical technology the foetus has become a patient. Nevertheless, this patient can only be approached through his mother, the pregnant woman. Not only in the sphere of medicine, but also in the sphere of morality this is a singular situation, since the woman's responsibility for the foetus cannot be substituted by any other person or institution. This is even the case when the foetus's survival *ex vivo* is principally possible. Therefore, it is not the status of a foetus as 'non-person' in the strict sense, nor the status of a human being with genetic disorders, nor the expected quality of life of the foetus (except in the case of low chance of survivability) that gives us the right to balance his or her rights but only the irreplaceable responsibility of the mother who may not consider herself able to take this responsibility. Put in general terms of moral tradition, a person does not act morally wrongly if it is practically impossible for him or her to fulfil the alleged duty. The reasons that practically (not theoretically!) repeal the moral claim cannot be stated in objective terms, but depend on the subjective view of the agent who nevertheless has to have reasons. It is in this way, that, on the one hand, morality encounters the good life-concept of a person, and, on the other hand, the good-life concept is crossing the moral claims.

In the context of prenatal diagnosis, we are faced with situations where pregnant women fear the risk of a genetic disorder causing a serious disease of the foetus. Since the protective responsibility for the foetus cannot be substituted by any other person, their subjective estimation of what they can bear has to be taken seriously not only for the woman's or couple's sake but for the sake of the future child as well. To judge this situation is difficult, however. For the woman's exclusive role is determined by the time of pregnancy, whereas the time after birth concerns herself *and* the father of the child. Therefore, the moral argument to give the woman priority in the decision can only apply for the time of pregnancy. Mostly, however, the woman's estimation will refer to the future, to the responsibility and possibly the burden of rearing a handicapped child. This estimation is not only subjective, it is also dependent on the parental relationship and the support the woman is getting, on social values and settings that allow her to integrate the extraordinary responsibility into her own life-plans, and it is dependent on the potential accuracy of the medical prognosis itself. All these uncertainties complicate the decision to undergo prenatal diagnosis and specifically invasive forms of prenatal diagnosis, since they have to be weighed against the life risk for the foetus that go along with these tests. Current procedures of non-invasive genetic tests conceal this moral conflict: they postpone it to the time of invasive tests, and even pre-testing counselling often disguises the ethical conflict that arises for those women who receive a bad result from amniocentesis or chorion villus sampling at a time where the decision for or against the child must be made more or less immediately.

Professional Responsibility

Ethical Guidelines

Physicians and genetic counsellors have to have ethical and, of course, juridical guidelines that form the normative framework of their actions. As the current directives were elaborated when prenatal diagnosis was considered to be an exceptional instrument of pregnancy monitoring, they have to be reconsidered in the present context where the exception has almost become normality. Therefore, a new evaluation has to take place, analyses of the development of the last decades debated, and, in my view, the moral status of the foetus reinterpreted. I have argued for general and more concrete considerations with respect to the protection of human foetuses. If these arguments are valid, the description of prenatal diagnosis as part of the private choice in reproduction is misleading. Furthermore, the principle of autonomy will have to be conceptualised as a moral principle that stresses the imperative of respecting others in their freedom and protecting those who can not yet (or no longer) live a life independent of our specific support.

In the background of these professional guidelines, some broader aspects of prenatal diagnosis have to be reflected: e.g. the medical and social context of the specific medical practice of genetic diagnosis, above all with regard to neonatology and paediatrics but also with regard to discussions about screenings and with regard to possible eugenic purposes. The consequences for the medical professionals – including the nursing staff – in their self-understanding have to be discussed according to the principles of medical action. The concept of genetic disorders as disease, the consequences for handicapped people, the financial resources for families, the financial resources for genetic counselling, the question of which institution should offer counselling and what education is needed, are to be reflected and discussed much more thoroughly than at present. Together they may give prenatal diagnosis and above all genetic counselling a basis that individual counsellors can work on.

Gen-ethical Counselling

According to a gen-ethical approach to counselling which reflects the counsellor's role as advocate of the woman as well as of the foetus, the current approaches of nondirective counselling must be revised.

It would have to be what I call gen-ethical counselling, with the professionals being not only the counsellors of the women/parents, but also the advocates of the foetus. Since the goal of prevention has been, and sometimes still is part of the professionals' view of counselling, some women or couples have been put under 'nondirective' pressure to undergo prenatal diagnosis. It could well be that many women need support of the counsellors in the other direction, namely in their respect and acceptance of a child that might be different from what they wish him or her to be. In my view, genetic counselling is justified only, if it is based on a concept referring to the moral protection and right of any human to come into existence. This normative framework is not questioned as such, if it is not possible to be practically applied in all cases. Following the reflections on the moral conflict between pregnant woman and foetus, there will be cases where imposing the moral duty to protect the foetus practically would lead to a kind of moral heroism that cannot be justified either. Thus, to have a general rule of protection does not mean that no context-sensitive judgment is needed. Quite on the contrary, context-sensitive judgment is part of applied ethical reasoning, and will finally be part of gen-ethical counselling.

According to this approach, the ethical dimension of prenatal diagnosis would have to become an integral part of the counselling process, and women/

parents could not be left alone in their decision but rather be supported in all phases of the decision-making process *and* the time after. It could even be necessary that the counselling institution should be moved out of the clinical institution and be established as a counselling institution of its own kind. This becomes evident when the following four-step model of gen-ethical counselling is considered.

First, the couple must receive all relevant information on prenatal diagnosis including that which deals with juridical, financial and ethically relevant aspects. This also applies to non-invasive tests as the triple test. Giving adequate information is a direct consequence of the principle of autonomy which gives the women and/or couples the chance to free and informed consent to each step of a medical procedure. At present, this step is at least ideally realised by physicians or counsellors.

Second, a special kind of ethical self-reflection should be pursued. The clients should be helped to outline their goals and purposes with respect to prenatal diagnosis. These can be considered in the context of personal life planning, ideals of one's identity and, in general terms, in the context of the good life of a person with respect to the wish for a child. By this the values and background of a person's ethical judgments can be reflected, and possible conflicts between the partners on the one hand, between patients and medical staff on the other hand, can be faced before genetic diagnosis has begun. This part of the counselling process has to be nondirective in its classical understanding, taking seriously the evaluative context of decision-making, without judging it in normative terms and excluding any manipulative directives with respect to individual life-plans. Since pregnancy is a time in which the identity concept is changing and the life-plan of a person is challenged by the uncertainty of the 'new' identity as mother or father of a child which will be very much dependent on his or her parents, a reflection on the basis of evaluative judgments is more or less taking place anyway. This is still true for further pregnancies. Women or couples who want to undergo prenatal diagnosis fear for their child and themselves to be carriers of a genetic disorder that can be detected. To be supported in the articulation of fears and possible conflicts means to take seriously the individual and subjective perspective which is based on personal experiences. It may be argued that this not the task of genetic counselling. It is, however, the task of gen-ethical counselling, because taking seriously the subjective perspective, the experience, emotions and hopes are part of the overall ethical reflection, traditionally expressed in an ethics of human happiness and virtue ethics.

In the third step the couple should be helped to reflect on the normatively ethical, i.e. moral problems, especially concerning the life of the foetus that might be at risk by the diagnostic procedure itself. Possible actions with respect to positive results of the diagnosis have to be discussed and considered in their moral relevance. In this part of the counselling process, counsellors have to take the advocative position for the foetus, in order to reveal his/her moral status and his/her moral rights. This is a direct consequence of their professional role that treats the foetus as a patient, and it stands in continuity with professional guidelines that set limits to prenatal diagnosis anyway. Therefore, to take an advocatory position on behalf of a foetus reflects not only the normative indications reflected above but also the professional ethics as part of the counselling process. Certain requests that have to be tolerated on the level of personal life-planning (step 2), will not be acceptable on the level of normative-moral reflection, i.e. the duty to respect another human being. For example, whereas it is understandable in certain contexts to request sex-selection without medical indication, it still is not morally tolerable to risk the life of a foetus for these reasons.

The last step of gen-ethical counselling is aimed at gathering the ethical and moral problems in a single perspective that finally results in a decision. Integration of the moral point of view into the personal ethical view of living one's life is the goal of this step. Within the limits of strictly interpreted medical indications counselling will be nondirective again. Integration of the moral responsibility for the foetus might not be possible in all contexts and situations, as indicated above. Therefore the counsellor's task will be to help the couple finding the solution that is closest to such integration. The possible compromises will be practical; they do not, however, touch the correctness of either the ethical or the moral judgments.

Gen-ethical counselling indicates that the decision-making process is so difficult on an individual level that clients should not be left alone with it. To offer them support in their actual decision may lead to the conclusion that long-term support in coping is needed. This may be the case if the result of the prenatal diagnosis is positive, no matter what decision is made by the woman or couple. Counsellors should be able to offer this support as well, either themselves or in delegation to specially trained therapeutic professionals.

In conclusion, gen-ethical counselling does not fail to appreciate the clients' autonomy to decide, but it takes seriously the moral decision that is involved in prenatal diagnosis. Its character remains nondirective insofar as any manipulation of the decision has to be excluded. However, to advocate the foetus as a patient and to make explicit his or her right to be protected, is in no means manipulative or, in this sense, directive, but is a consequence of the reflection of the moral responsibility of the professional staff itself. In genethical counselling, this shared responsibility between client and counsellor has to be made explicit. Even though the last and final parental responsibility to decide what action is best for the foetus *and* the family cannot anticipated by the professionals, they are partly responsible for the actions, too. Professional guidelines that set the limits for the principal legitimacy of prenatal diagnosis explicate the framework of this responsibility. The individual counselling sessions have to concretise and, thereby, contextualise the moral decision that is involved in prenatal diagnosis. Both framework and directives of genetic counselling have to be strongly dependent on ethical evaluations to establish a practice that is not only legally but also morally justified. As I have argued the analysis of the concept of personhood gives us strong reasons to ascribe dignity and moral rights to foetuses who may not be entirely instrumentalised for the purposes of other persons and their rights. This right is not to be balanced with the right of a person itself, but only with the estimated practical incapacity to take a responsibility for a child dependent on its mother or parents. Where this inability is foreseen before procreation, refraining from having children might show more moral responsibility than chancing the conflict of terminating a pregnancy. On the other hand, when a commonly offered medical diagnostic practice implicitly suggests that a morally acceptable solution is available in a late stage of pregnancy, this practice comes into conflict with the medical task itself.[6]

Notes

1 Since foetal blood sampling will not replace invasive techniques in the near future, it is not considered with respect to the present practice of prenatal genetic testing.
2 See also pilot studies in Denmark and the United Kingdom, Schwartz, 1993 and Mennie, 1992.
3 The risks include termination of pregnancy, statistically estimated between 0.3–1 per cent for amniocentesis, even higher for chorionic villus sampling, and the risk of injury to the foetus depending on the experience of the physician.
4 Cf. Rogers, 1972; Reed, 1974; Clarke, 1994; and Bartels, 1993.
5 Only in the relationship of moral agents the Golden Rule appears to be too narrow, it is necessary to respect the other in *their* wishes and needs that my differ from ours.
6 This paper is the expanded version of a previous article 'Ethical Responsibility in Prenatal Genetic Diagnosis', *Biomedical Ethics*, Vol. 2, No. 3 (1997), pp. 78–85.

References

Baitsch, H. and Reif, M. (1986), *Genetische Beratung. Hilfestellung für eine selbstverantwortliche Entscheidung?*, Berlin/Heidelberg.

Bartels, D.M., LeRoy, B.S. and Caplan, A.L. (eds) (1993), *Prescribing our Future. Ethical Challenges in Genetic Counselling*, New York.

Birnbacher, D. (1993), 'Ethische Probleme der Pränataldiagnostik aus der Sicht eines Philosophen' in B. Schöne-Seifert and L. Krüger (eds), *Humangenetik – Ethische Probleme der Beratung, Diagnostik und Forschung (Jahrbuch des Arbeitskreises Medizinischer Ethik-Kommissionen in der BRD)*, Stuttgart.

Clarke, A. (ed.) (1994), *Genetic Counselling: Practice and Principles*, London.

Gewirth, A. (1978), *Reason and Morality*, Chicago.

Habermas, J. (1988), 'Individuierung durch Vergesellschaftung. Zu G. H. Meads Theorie der Subjektivität' in *Ders.: Nachmetaphysisches Denken*, Frankfurt am Main.

Haker, H. (1997), 'Ethical Responsibility in Prenatal Genetic Diagnosis', *Biomedical Ethics*, Vol. 2, No. 3, pp. 78–85.

Haker, H. (1993), 'Human Genome Analysis and Eugenics' in H. Haker, R. Hearn and K. Steigleder (eds), *Ethics of Human Genome Analyis. European Perspectives*, Tübingen.

Haker, H. (1998), *Moralische Identität. Literarische Lebensgeschichten als Medium ethischer Reflexion. Mit einer Interpretation der 'Jahrestage' von Uwe Johnson*, Tübingen.

Harris, J.C. (1991), 'Aborting Abnormal Fetuses: The Parental Perspective', *Journal of Applied Philosophy*, 8, pp. 57–68.

Honneth, A. (1992), *Der Kampf um Anerkennung. Zur moralischen Grammatik sozialer Konflikte*, Frankfurt am Main.

Kant, I. (1956), *Grundlegung zur Metaphysik der Sitten* (Werke in 10 Bänden, hg. v. W. Weischedel, Bd. 6), Darmstadt.

Locke, J. (1981), *An Essay Concerning Human Understanding*, Dt.: Über den Verstand, 2 Bde, hg. v. R. Brandt, Hamburg.

Mennie, M. et al. (1992), 'Prenatal Screening for Cystic Fibrosis', *The Lancet*, 339/340, pp. 214–16.

Reed, S.C. (1974), 'A Short History of Genetic Counselling', *Social Biology*, 21, pp. 332–9.

Ricoeur, P. (1990), *Soi-mème Comme un Autre*, Paris.

Rogers, C.R. (1972), *Die nicht-direktive Beratung*, München.

Schöne-Seifert, B. and Krüger, L. (1993), 'Humangenetik heute: umstrittene ethische Grundfragen' in B. Schöne-Seifert and L. Krüger (eds), *Humangenetik – Ethische Probleme der Beratung, Diagnostik und Forschung (Jahrbuch des Arbeitskreises Medizinischer Ethik-Kommissionen in der BRD)*, Stuttgart.

Schroeder-Kurth, T. (1993), 'Ethical Standards of Genetic Counselling' in H. Haker, R. Hearn and K. Steigleder (eds), *Ethics of Human Genome Analyis. European Perspectives*, Tübingen.

Schwartz, M. et al. (1993), 'Screening for Carriers of Cystic Fibrosis among Pregnant Women: A Pilot Study', *European Journal of Human Genetics*, 1/3, pp. 239–44.

Witt, D.R. et al. (1996), 'Cystic Fibrosis Heterozygote Screening in 5,161 Pregnant Women', *American Journal of Human Genetics*, 58/4, pp. 823–35.

Wolff, G. and Jung, C. (1994), 'Nichtdirektivität und genetische Beratung', *Medizinische Genetik*, 6, pp. 195–204.

8 Embryos *in vitro*, Personhood, and Rights

MAUREEN JUNKER-KENNY

At a time when the boundaries of human life are moving just as rapidly as cultural values are shifting, one could be forgiven for thinking that genetics is being transformed from an expert culture to a life form. It comes to shape not only our lifeworld but the way we think of ourselves. In a scientific era which has to defend the sense of personal agency even of its adult citizens over against cultural, gender, and genetic determinations, it seems difficult to base a claim for personal dignity on the potentiality of developing human life. Yet the challenge some understandings of genetics pose to our every day understanding of ourselves as capable of reflection and action is linked to the issue of how we conceptualise personhood. Can it simply be attributed to human beings in the full power of their faculties, might it only exist in peak performances when we find no reason to distance ourselves from our actions and motives? Or can an element of growth and humble beginnings be admitted, possibly even interludes and lapses that are less than fully mature and dignified? Does personal dignity therefore always remain a matter of struggle and mutual encouragement, and is autonomy not a stage reached once and for all but the fruit of creative anticipation that remains in need of confirmation? If these assumptions are true, then the analysis of positions on the status of the human embryo might not just be reflections on a distant matter of our concern or on our remote beginnings but offer insights into fundamental realities of the human constitution.

For many years, the primary context in which the moral status of the human embryo was discussed, has been the question of pregnancy termination. While an analysis of the controversial positions taken up in this debate can still help identify the different possible starting points, relevant arguments and conclusions, the decisive difference between the contexts of crisis pregnancies and the new reproductive and genetic techniques needs to be highlighted. Before I consider different argumentations for and against the personhood or person status of the human embryo, I shall first deal with the way in which the location of this embryo, *in vivo* or *in vitro*, affects moral judgment. In my third section, I shall explore how the issue of 'right' and the adequacy of 'rights language' have to be assessed on the backdrop of cultural changes affecting the lifeworld through individualisation and market imperatives. The fourth section concludes with the impact of the new reproductive and cloning techniques on theology and the style of argumentation appropriate for theological ethics to engage in public interdisciplinary debate.

Embryos *in vitro*

The idea that the physical location of an entity could make a difference to the philosophical determination of its nature and properties, must sound utterly suspicious to a mind trained in Western systems of classification. Unless one wanted to make a postmodern point about the arbitrariness of logical orders imposed upon ungraspable forces of life, one had better steer clear of such untenable claims. However, what is at stake with the difference between *in vivo* and *in vitro*, is not so much a question of locality but of natural processes over against technical procedures; it is the step from natural chance to the purpose-oriented action of the human will, from spontaneous development to the power to design, evaluate and possibly reject. Also 'planned' and hoped-for pregnancies come to be as a matter of chance while all of them, not just those that pose a conflict, at the same time signify more than a natural process for women and include taking a stance towards the fact of a new life's beginning within their own. Yet, the crucial moral difference between the *in vivo* and *in vitro* locations is that in the latter, a situation in which the embryo is exposed to possible manipulation is set up as a matter of rational planning, investigation, and perfection, and justified through the goal of helping infertile couples to have a child of their own. The techniques employed to assist this desire therefore have to be oriented by the standard of personal dignity in order not to compromise their goal. This criterion becomes all the more decisive since with the growing technical skills and insights, e.g. in the frame of the Human Genome Project, the power difference between the biotechnological experts and their objects is increasing at an exponential rate. The opportunity for and degree of exploiting and instrumentalising embryos for the purposes of third parties mount accordingly. It is this 'context of the scientific manipulation of embryos' which urges the French philosopher Paul Ricoeur to call for special caution in the philosophical determination of the status of the embryo (Ricoeur, 1992, pp. 272–3). This determination arises from a 'complex play between science and wisdom' (ibid., p. 272), in which the faculty of judgment is given the function to call science to stand back in cases of doubt.

The ethical dilemmas arising from a pregnancy that is unwanted, resulting from rape or problematic for other reasons, are situated differently in that they do not form part of planned, rationally organised events. People find themselves in them and unexpectedly, are called upon to act. The balancing of the rights of everyone affected and of the norms and goods to be taken into consideration takes place in a context of unintended emergency. While the standard of personal dignity remains the same in both cases, and the concept of responsibility encompasses more than an agent's intentions, there can be cases of conflict between the rights of the pregnant woman and those of the embryo which can only be called tragic. This evaluation, however, would clearly be inappropriate with regard to technically planned processes since they have to figure expenses, consequences and losses into their calculation of success ahead of implementing them. On this point, Klaus Steigleder argues in a similar way:

> Things are quite different [from abortion in a conflict situation] if the conflict is created in a controllable and somewhat institutionalized way as a side effect of a medical procedure ... 'pregnancy reduction' may be a morally tolerable solution in unforeseen tragic situations, but it is not tolerable in the context of a knowingly induced, though unintended, problem (Steigleder, 1997, p. 245).

Even if the status of the embryo is uncertain, the evidence of its exposure and lack of protection *in vitro* should call forth express alert and determination to resist the chance to exploit this condition. Yet, it is precisely the clinical objectification made possible through *in vitro* techniques that endangers an attitude of respect for human individual life and lends itself to a mentality of control, if not experimentation. In the laboratory situation, the typical conflicts of a medical doctor who is at the same time a researcher, arise to a heightened degree. She has to balance the good of her present patient with the hoped-for insights into disorders that will be of benefit to future patients. Each time she encounters her actual patient who is recognisably a fellow-human being, she is reminded of the limit each person poses to the interests of others, even future fellow-sufferers. Medical and laboratory staff in *in vitro* procedures merely encounter clusters of cells which only their knowledge and imagination may help to envision as future children. The conflict of interests is easily tilted in favour of future insights at the expense of the present 'patients' whose value, moreover, is derived from the intense wish of their parents for a child of their own. The difference to an embryo *in vitro* is that *in vivo*, a personal relationship exists even if it is one of struggle. The future child that the embryo will develop into, evokes feelings such as happiness, doubt, or despair. While an embryo *in vitro* will represent the parents' highest hopes, the experts dealing with the contents of petri dishes will pursue methods of best practice by observing the division rate of cells and selecting the most prospective candidates for implantation. The impact of this objectifying method of science – although in the service of a couple's wish for a mutual child – on how we interpret the beginnings of human life, should be noted.

Personhood in Anthropological Perspective

Before presenting different approaches towards personhood and to the problem of whether it can be attributed to embryos, before then going on to evaluate them on how much they take anthropological considerations seriously, I shall first clarify the philosophical and interpretive nature of this enterprise, in other words, its hermeneutical character.

Defining the Beginning of Human Personhood: Practical Hermeneutical Circle and Descriptive Evidence

Any definition of the beginning (and end) of human personhood is caught in a practical hermeneutical circle. We define its starting point because we want to act in a certain way, and we act according to how we have defined it (Mieth, 1990, pp.78–82, 95–6). A similar evaluation can be found in Paul Ricoeur's discussion of 'respect for persons at the "beginning of life"'. In Kant's

> bipolar opposition of persons and things, the distinction between mode of beings remained inseparable from practice, that is, from the manner of treating persons and things … To be sure, the identification of thresholds and degrees marking the appearance of properties of being is dependent on science alone. But the ontological tenor assigned to the predicate 'potential' in the expression 'potential human person' is perhaps not separable from the manner of 'treating' beings corresponding to these various stages. Manner of being and manner of treating would seem to be mutually determined in the formation of prudential judgments occasioned by each advance in the power that technology confers today on humankind over life in its beginnings (Ricoeur, 1992, pp. 270, 272).

In practical terms, if we consider, e.g., the moment of implantation in the uterus, or the presence of brain activity or the ability to communicate as the starting-point for ascribing personhood, we are free to treat the embryo prior to this stage in any way we consider useful.

On the other hand, one's practical interest has to be supported by descriptive evidence if it is not to be judged arbitrary and decisionist. Morally relevant are those features which relate to the human being's destination to be free, her capacity to be a moral subject or agent. From the circular structure of the definition two tasks arise: one, to justify the guiding perspective which informs one's proposition; second, to base the attribution of human dignity or personhood on communicable reasons.

Each definition therefore also has a practical intent. Once we ascribe human life and personhood to an entity, we want to protect it. A maximal definition of human life, such as the ability to communicate, or to act independently, offers minimal protection to the stages prior to these competencies and after they have been lost. If one wants to give maximum protection, one has to use a minimal definition, such as the new auto-reproductive genetic unit created by the fusion of the nuclei of egg and sperm. To ensure the protection of 'the weakest link in the chain of the human species' (Folscheid, 1996, p. 21) at the earliest possible stage, one would regard personhood as beginning with syngamy.

The hermeneutical insight into the practical circle operative in defining the beginning and end of human life and personhood, undermines two positions which have frequently marked the opposite two poles in the debate on the moral status of the embryo. One is a philosophical essentialism according to which the embryo 'is' a person. The other is marked by an equally essentialist empiricism which either measures personhood in a purely actualistic way from attestable qualities such as consciousness, or else translates thresholds of biological development immediately into degrees of becoming a person and stages of greater value. The circumstance that human observation can analyse (and human interference disrupt) an integral process of development cannot be taken as evidence that the sequences we discover are in fact separate stages on which different degrees of value can be founded.

The statements of 'facts' offered by both sides are not acceptable from a hermeneutically enlightened perspective. One cannot abstract from the interpretive task and the insight that the respective definition will determine one's action and one's practical interest the definition. Awareness of context and consequences are constitutive for the hermeneutical view. This admission, however, does not necessarily lead to an invitation to a pragmatism which would permit all kinds of interference with developing human life, since nothing objective can be said on the status of the embryo. The realisation that practical options are involved in these theoretical matters can equally sharpen one's sense of responsibility.

Philosophical Positions on Personhood

If the new discipline of biomedical ethics does not want to be restricted to offering *post factum* justifications or criticisms of each new scientific breakthrough, thereby acquiescing to the pace of biomedical development, it has to develop categories in advance by which to guide and assess justifiable directions of enquiry and priorities of research. It has to measure its recommendations not by technical feasibility but by anthropological assumptions and considerations of the impact of biomedical science on the lifeworld. Therefore, it cannot leave reflections on the human constitution aside, such as its synthesis of autonomy and facticity, the quest for rightness and fallibility, and the Kantian insight that it is a mark of the dignity of human life that it cannot be evaluated. A discipline at the cutting edge of future technologies and progress in making human life plannable, controllable, and rationalisable in its beginning and end, needs to be able to interpret the core and the limits of human existence, and the meaning of long-term cultural achievements in the history of humanity. Guiding ideas in Western constitutions such as inviolable and irreproducible identity, equality, solidarity with the weak, and recognition of children in their own right, were developed in intellectual and political struggles and represent an historical achievement that cannot be taken for granted and that has to be defended against new contemporary challenges.

The positions presented in the following converge in the interest to attribute certain rights of protection to the embryo in view of its uncertain status while they differ in the scope of the rights ascribed and in the goals and criteria of justification. They are both located on the opposite side to an argumentation such as Peter Singer's for whom being human is not a necessary condition of being a person, since in his view personhood depends on actual and empirically verifiable features like intelligent behaviour. On that count, some animals perform better than human newborns. To ascribe personhood to these just because they are human amounts to the famous fallacy of 'speciesism', an unjust partiality to one's own which he judges to be on a par with racism and sexism. His 'practical ethics' can be counted as a strategy to rid society of morality, as a factor of repealing the existing minimal consensus that is the last remaining common ground within a pluralist society (Wils, 1993, p. 10).

The two following argumentations therefore have not been chosen to represent the most extreme views in this field but to highlight the different elements and their respective weights in attributing personhood and personal dignity.

Dignity Based on Ontology

The first approach I want to discuss seems to fulfil all the demands for an anthropological reflection which also takes into account natural features, inclinations, and basic 'givens' of human subjectivity. Ludger Honnefelder wants to show 'that it is the inner context of two aspects which one can call the "person principle" and the "nature principle" which are at the core of the ethos of human rights and furnish the foundation of a consensus of modern society' (Honnefelder, 1996a, p. 215).

In his attempt to ground an 'absolute' over against a 'relative' entitlement for embryos to be protected (1996b, p. 2), he claims Kant for his interpretation that a substantive ontology has to be postulated.

> Kant decides without hesitation in favour of regarding personhood and human nature as co-extensive (*deckungsgleich*). At the same time, however, he has to claim a unity of both which according to his premises theoretical reason cannot know, and has to claim a concept of substance in the practical sphere which cannot be demonstrated in the theoretical sphere (Honnefelder, 1996a, p. 238).

Apart from the question whether it is a concept of substance that Kant claims, two critical points have to be raised.

One decisive distinction is obscured in the following argumentation:

> Person principle and nature principle are indissolubly linked. The capability (*Vermögen*) stated in the concept of the person to be a subject has to be considered as the ground for the attribution of human dignity; to be a living being of a certain nature has to be considered as the criterion for the attribution of the property stated in the concept of the human being (ibid., pp. 260–1).

What Honnefelder calls a 'criterion', the feature of being a bodily human being, only refers to the problem of knowing and is not the criterion of validity (*Geltungskriterium*). If it was, then corporeality would be a normative reason for human dignity. This empirical 'criterion' only answers the question of knowability (*Erkennbarkeit*) who is a free human subject, namely whoever has human shape, but is not a criterion in the normative sense.

The second problematic step is his extension of the two interlinked principles of person and nature to embryos who do not have a recognizable human shape. This extension is based on a doubtful explanation of the continuity between the zygote and the future person in terms of substance metaphysics. Honnefelder sees the relation between the zygote and the future person in terms of a natural teleology, rather than as physical, temporal and historical continuity.

Therefore, it can be questioned whether his attempt to combine the 'person principle (being a moral subject)' with the 'nature principle (belonging to the human species)' is successful. His terms mirror the teleology of substance metaphysics:

> The embryo is a person because it is a human being, and the human being includes the development form of the subject ... Human beings are entitled to person status because by nature they possess the capability to be subjects (Honnefelder, 1996b, p. 10).

Honnefelder's ultimately naturalistic concept of the person cannot do justice to the evidence that embryos have to develop into free moral subjects. The tradition of substance metaphysics seems to be particularly badly equipped to conceptualise a life whose being – to quote a famous theological title, Eberhard Jüngel's *Gottes Sein ist im Werden* – consists in becoming, namely a self-reflective subject. Honnefelder's attempt to mediate between the Kantian foundation of dignity on the capability to be a moral subject and an Aristotelian natural teleology comes down on the side of natural law essentialism. Is it possible to understand the embryo's potentiality not in terms of a substantial development form, but in terms of a philosophy of recognition?

Dignity Based on Moral Agency

In his explanation of Alan Gewirth's foundation of human rights on the mutual recognition of subjects with intentions for agency, Klaus Steigleder shows that for Gewirth the reason for personal dignity is the actual capacity for agency which at the same time claims the recognition of the other. How can this argumentation be applied to beings who are only potentially capable of agency?

> The basis of dignity is that the agent inevitably has purposes she wants to fulfill ... because agents must attribute to themselves and to each other dignity they must, in different ways, regard human non-persons as connected to their dignity

> and therefore must confer, in different ways, moral status upon them (Steigleder, 1997, pp. 241, 239).

The reason for their derived moral status is their potential agency.

> [H]uman beings who are not yet agents must possess moral significance for agents for the sufficient reason that they possess the potentiality to become agents ... They do not possess the same moral status as agents for they do not possess dignity (ibid., p. 241).

Yet their status gives them extensive rights at least in principle:

> The general normative implication of the moral status of human embryos and fetuses as potential agents is a general prescription to preserve, protect and foster them. This implies the general prohibition of harming and killing them (ibid., p. 244).

However, the duty to protect embryos as potential agents according to Steigleder does not apply to the stage of the so-called 'pre-embryo' before the stage of implantation. He does not recognise any categorical reasons against their being used in wasting research. He distinguishes between two potentialities with different normative implications. The

> above argument from potentiality refers to the embryo in the strict sense and not to the so-called 'pre-embryo'. By this term is usually meant the fertilized human egg (zygote) and the further stages of its development during roughly the first two weeks ... The pre-embryo has the 'potentiality to produce', for instance, the embryo. The embryo itself has the 'potentiality to become' an agent. It was this last kind of potentiality which figured in the argument from potentiality. Thus, the moral status of the human pre-embryo is different and much weaker than that of the human embryo. It is difficult to see how the status of the pre-embryo as such could justify a categorical prohibition of its use for research. On the other hand, the question is to be considered to what extent and with what consequences we have to attribute to the human pre-embryo symbolic significance (ibid., p. 245).

The main point to be made against separating the 'pre-embryo' from the 'embryo' and situating the fertilized egg closer to the gametes before fertilisation than to the implanted embryo, is that it plays down the physical and temporal continuity between the zygote, the implanting embryo, and the foetus. To assign to the 'pre-embryo' the activity of 'producing' and to the 'embryo' one of 'becoming' undervalues the fact that the 'pre-embryo' itself partly becomes the embryo, partly produces the placenta. Steigleder only speaks of the 'embryo' after its differentiation from the placenta and considers the zygote as the precondition for the embryo but not as in part identical with it. This is why the only value it can have before implantation is a symbolical one.

The point to be pursued in our context, however, is how the criteriological value of the anthropological presuppositions for agency is revised. In an earlier article Steigleder had pointed out the features shared by full agents, embryos and pre-embryos, and had attributed higher moral significance to them. The 'context of bodily constitution and the accompanying historicity of the agent' (Steigleder, 1993, p. 119) figured as additional dimensions besides the potency for agency in attributing dignity. Especially the element of historicity

> undermines the possibility of an agent to measure consistently the limits of dignity in an abstract 'Either' (agent) – 'Or' (non-agent). It points to the lines of development and continuities beyond agency and does not permit any 'neutrality' with regard to the neediness given with corporeality which represents an essential element in the agent's justification of his rights claims. Moreover, it is due to the bodily constitution of someone capable of action that he recognizes himself in human beings not capable of action . . . This is why they do have a morally relevant status after all which is the basis of obligations. This is why treating human beings not capable of action in a different way needs to be justified (ibid., p. 120).

Instead of postulating discontinuity, the less restrictive earlier version sees ethical implications in the features zygotes share with agents: bodily constitution and historicity. Both the earlier and the later positions, however, come to similar conclusions in which embryos cannot figure in categorical argumentations which can only apply to agents in the full sense. They both find evidence pointing to the conclusion that

> human beings not capable of agency ... participate in certain respects in the dignity of subjects capable of agency. Insofar the concept of a human dignity which extends farther than personal dignity makes sense (ibid., p. 121).

However, the earlier version takes something into account the lack of which can later be seen as the principal weakness of an ethical approach which attributes personal dignity only to moral agents: its neglect of the need to base ethics on a theory of the human being, instead of first narrowing moral theory down to the problem of justification and then deriving applications from this lean foundation. While the ground of obligation to treat each other as ends-in-ourselves and never simply as means lies in the unconditional character of freedom, this autonomy as the basis of justification implies a whole anthropology. 'Lines of development and continuities beyond agency' (ibid., p. 121) need to be explored first if one does not want to succumb to what Dietmar Mieth has called an 'ethicistic fallacy' (Mieth, 1998, pp. 51, 56–9). This ethicistic reduction is evident in Steigleder's warning not to lose sight of the true relations of justification and derivation:

> In concrete ethical argumentation, however, the relations of justification and derivation have to be kept clear. Under no circumstances is it valid simply to pursue 'conceptual' argumentations – e.g., to conclude from the new human life which the fertilized egg cell represents, and 'human dignity', a moral status of the zygote which would be comparable to that of a subject capable of agency (Steigleder, 1993, p. 121).

While the methodological critique may be valid, it is itself based on too slim a foundation. The problem with this portrayal of the logical order of justification is that it argues from the level of 'ought' to the level of 'is' and assumes that the rights justified for full agents cannot extend to those who are only potential agents. The opposite order of argumentation would begin with the anthropological presuppositions of freedom, of which agency is an expression, and acknowledge them as morally relevant conditions for developing autonomy. Bodily constitution, temporality and historicity are not the normative ground for recognising the other, but since the subject of recognition does not come without them it would be hard to imagine how humans could respect and promote each other's freedom apart from their embodiment. The fact that autonomy takes a long time to be fully realised then cannot be counted against protecting those on the way towards it. In his philosophical reflection on organic development Christian Kummer (1997, p. 16) puts this point in the following way:

> Should all the stages before be ethically irrelevant only because we do not know anything about how much delay the autonomy of the human process of development can afford under the protection of the maternal organs of supply?

It therefore makes much more sense to honour and foster human life as a 'fundamental and integral value' (Mieth, 1990, p. 100). 'Integral' means that it is impossible to distinguish between a stage in which human life does not have any value and is at the disposal of our purposes, and a later stage where it acquires full value.

The utilitarian ethicist Birnbacher supports this judgment for reasons of ethical clarity and consistency. Although he does not consider wasting research on human zygotes and embryos to be 'absolutely morally unpermissible' (Birnbacher, 1987, p. 86), since for him 'the existence of the human being as an independent individual, as a subject' who has individual human dignity only begins at birth (ibid., p. 80), he warns:

> It would mean a confusing incongruity for our moral sensibility to treat the pre-personal and pre-individual pre-stages which stand in full continuity to a fully formed personal existence unscrupulously as a thing while we accord personal existence the full amount of protection which, among others, respect for individual human dignity requires (ibid., pp. 85–6).

Human life as an integral value is defended as well by Reiner Wimmer whose distinction between personal dignity and person status could be acceptable also to Steigleder's position. He reserves the term 'personal *dignity*' for actual agents while establishing comprehensive rights to protection on the potentiality of developing human life. Being destined to become a person, it has to be ascribed person *status*.

> I would like to distinguish between attribution and denial of personhood and attribution and denial of the [legal or moral] status of a person, because there might be humans who are not persons but are endowed with legal or moral rights, so that an inference can be drawn from being a person to being in possession of the status of a person but not the reverse. Although it is illegitimate to calibrate personhood and the respect due to it, there are situations in which it is clear that human beings do not, or no longer, or not yet, possess personhood. That is the case with deceased human beings or with foetuses. However, the status of persons is ascribed to them, and this entails the relevant practical consequences, e.g. … to ban experiments … The distinction between personhood and the status as a person … presupposes that, while the personhood of a human being has a beginning and an end in time, this is not the case with its status as a person. Another presupposition is that it makes sense to speak of the status of person only if a living being is generically … able to become a person. This avoids making the ascription of the status as a person dependent on whether human beings are *willing* to grant one of their descendants this status or to provide the chance for him or her to become a person … the duty to have this kind of respect, as it is formulated in the moral person-principle, relates not only to existing persons but also to not yet existing but under normal conditions

> – i.e during an undisturbed development – developing persons (Wimmer, 1997, p. 89).

> Today there is a tendency to treat the concept of person not as an experiential concept ... but as an empirical one in the sense of the sciences: the status of person allows grades, somebody is more or less of a person so that he or she as a person is measurable by degrees so that we would be able to identify higher and lower levels of personhood and consequently to attribute different values to different developmental states of personal beings and to weigh between them in situations of conflict. The traditional view according to which all humans are persons in the same way ought to be differentiated – so at least the opinion of some bioethicists –: otherwise we would ignore the obvious fact that there is a large, morally relevant difference between a human foetus and a healthy adult with respect to their personhood and their status as a person. I think that the very point of the classical conception of personhood is not to allow any gradation so that the real differences synchronically between individuals and diachronically between various states of one and the same individual are irrelevant concerning the ascription of personhood; they are highly relevant concerning the treatment of a person namely in accord with his or her needs in a particular situation (Wimmer, 1998, p. 6).

A Third Approach: Dignity Based on Being Destined Towards Freedom

Between the ontologising approach based on the teleological potentiality of nature and the ethicistic position of founding human dignity on agency, a third kind of categorical argumentation seems possible which includes also *potential freedom* into the ethics of recognition. It grounds human dignity neither on a given human nature nor on actual performance but on the destination for freedom (Junker-Kenny, 1998b and 1999). Kant's concept of human dignity based on the capability to be moral can be extended to human beings who are presently not yet or no longer able to give evidence of their moral capacity. The anthropological conditions of freedom are taken seriously. It is a matter of everyday ethical experience that freedom only comes into existence if it is recognised in advance. It has to be anticipated in order to become real. The recognition of the other fundamentally has an innovative and anticipatory dimension and often enough has to be vicarious even among equally able agents. Once the basically advocatory nature of everyday ethical praxis has become clear, it is only a question of gradation not of principle to include humans who are not yet or no longer agents. Everybody's freedom needs the free recognition of others to be mediated to itself (Pröpper, 1995). Thus, it is possible to maintain that the capability for being a moral subject is the normative ground for personal dignity, yet not to use this criterion for excluding those forms of human life from protection which actually, cannot (yet) meet this claim. Whoever has the potential to be a free subject, is included in the categorical imperative of recognising the other as the supreme limit to one's interests.

Wimmer equally accords high significance to the anticipatory structure of realising personhood. Besides the historical continuity and the adult consciousness of retrospective identity with the embryos we were, this is the third factor for giving embryos protective person status.

> There is not only the continuity of history of a living being but also the conception of perdurance of personal identity during all stages of development the beginning of which is the … process of conception as the beginning of a whole though not yet fully individualised being, as embryology tells us. The phase or event ontology of human beings presented for instance by Derek Parfit and Peter Singer neglects that aspect of identity conscience and is therefore in discord with our self-understanding and self-estimation.
>
> In my opinion the last mentioned ascription of the status of person can be justified not only in retrospect out of the self-understanding of an adult but also by reference to the assumption that a human being without the requisite abilities of a person – … a human embryo, foetus and new-born – has in normal circumstances … the capacity or the capability of becoming a person in the sense of acquiring and executing the relevant abilities. The attribution of the status of person in that third respect is, theoretically speaking, a contrafactual presupposition because the genetic and physiological outfit alone is not sufficient for becoming a person. But it is at the same time a practically effective anticipation without which there would be no human persons: Human parents have to treat their new-borns as persons (as if they were already persons) in order that their children become persons … Even when parents plan to have a child which is not yet procreated, they already enter into moral obligations towards it. Thus, they treat it anticipatorily and imaginatively as a person, in so far as they have to practically respect all those demands of the child which they have to fulfill (Wimmer, 1998, pp. 8–9).

Lifeworld and Solidarity, Individualisation and Market Imperatives

Both 'anthropology' and 'lifeworld' serve as shorthand terms to indicate normative, historical and cultural dimensions and thus counteract reductive and overly abstract notions of the human person. From the normative implications of anthropology to the cultural notion of the lifeworld is only a short step. While the dangers of this concept have been pointed out – romanticising the lifeworld, claiming immediacy for it, and a tendency to 'naturalise' ethics (cf. Henrich's critique of Habermas, 1987), they do not preclude its usefulness. 'Lifeworld' as the shared background assumptions and practices of a community highlights the fact that the attitudes and expectations, the rites and legal regulations shaping and ordering first-order relationships – e.g. the recognition of childhood, equality and mutual commitment of both partners, respect and care for the elderly – represent historical, cultural, and humane achievements that took a long time to develop and that cannot be taken for granted.

The moral philosopher Brenda Almond (1998) points out

> the deep importance of human kin connections from both a personal and community perspective (ibid., p. 6) ... Family bonds are the cement of social existence and are not subject to construction and destruction by something as volatile as individual choice (ibid., p. 4).

She warns of a

> developing social consensus, led by some utilitarian philosophers as well as some feminist theorists, that technology having brought us into a new age of relationships by choice, this is something to be unproblematically welcomed. But this is a remarkable assumption in view of the history of family bonds in the story of the human race ... until now, birth and procreation and the inheritance of genetic material have been fixed landmarks in human culture and experience. Indeed, the changes that have been made possible by new technologies ... cannot negate the essential naturalness ... of conception, pregnancy and childbirth, the relationships they create and those on which they depend ... The contrasting claims – that biological parenthood is negligible – has wide implications, with more and more unpalatable procedures being seen as unproblematic: harvesting viable eggs from a foetus, for example, implanting eggs in women in their late fifties or sixties, casting corpses, male or female, in the role of parent – none of these with any real regard for the consequences, psychological and emotional for the resulting human beings. And yet consequences there are likely to be in many such cases. For such children are born orphans in a hitherto unknown sense – they are born already exiles from the kinship network (ibid., pp. 4–5).

Theorists of late or reflexive modernity such as Ulrich Beck have highlighted in what way primary relationships are under pressure not only from new biomedical techniques but from the general sociological and economic trend towards individualisation. While liberating the enterprising individual from the constraints of social standing and gender roles, the network of family and neighbourhood relations which formerly supported also 'unproductive' members, is undermined by the need for individuals to comply with all the exigencies of the labour market, e.g. competitiveness, mobility, extended work hours, and to subordinate the other segments of their life to this primary goal. In the context of this article, the focus is not on the psychological strains on those subscribing to these imperatives, or on the 'losers of modernity' who have fallen out of the system, and those who lose a whole world of social interaction if they are forced to leave a neighbourhood. The question here is, how does it affect the experience of childhood when the codes which generations of interaction have built up as their lifeworld are now subjected to the planning rationality of experts providing individualised 'head starts' even before birth? What change in attitude does it signify when a lifeworld that used to respect children in their own right and at their own rate of development becomes a laboratory for perfecting human beings like raw material in need of being processed? Regine Ammicht-Quinn warns that the short period in modernity when children were no longer instrumentalised for the economic survival and old age interests of their parents might already be drawing to a close. With the new possibilities of modelling children in their most elemental genetic constitution according to the desires of adults, an unprecedented extent and radicality of control determines their relationship. Unconditional acceptance might not always have been a reality for children, but the objectifying attitudes of quality control and legal redress change what should be a relationship into the trial of a consumer testing products.

Two recent articles criticise the cultural conflict evident in such choices with categories of alienation and of the antagonism between the market and the constitutional views of the human person. Kathryn Morgan's feminist rights critique and Dietmar Mieth's analysis of the struggle between fundamental rights and economic interests provide tools for assessing the contemporary cultural situation. At what level can the crisis of values evident in the present developments and in the reactions to them, be addressed effectively, and what is the significance of people's ultimate assumptions?

A Feminist Rights Critique

The adequacy of 'rights language' has been questioned from different perspectives, e.g., neo-Aristotelians, pro-modernity communitarians, and liberal theorists of radical democracy. With a view to scarce resources, there are limits to conceptualising desires as 'rights'. Inappropriate 'rights' language can promote the further encroachment of the 'colonising' forces of the legal and political systems on the first order and second order relationships fostered in the 'lifeworld'. Rights claims have to be analysed on the backdrop of cultural changes affecting society through individualisation, market imperatives and the dedication to promote instant gratification.

Kathryn Pauly Morgan explores the possibility that some 'rights' uncritically service 'needs' that have only been created by the existence of market-oriented options and facilities. She sees cosmetic surgery and techniques of assisted reproduction converging in the technological imperative dominating both of them. Not the genuine wish for a child is problematic, but the double face of the new reproductive technologies (NRT) which attach multiple conditions to their use. She highlights the ambivalence of the range at hand by identifying 'three paradoxes of choice'. They include, '1. The Choice of Conformity: Aspiring to Be Number "10", Aspiring to Have "Perfect" Babies. Paradox 2. Choosing Dependence in the Name of Autonomy. Paradox 3. Coerced Voluntariness and the Technological Imperative' (Morgan, 1996, pp. 226–33). An enlarged scope of choice is combined with a new degree of pressure to undergo the tests and techniques on 'offer'; the technological revaluation of the 'natural' as the 'primitive' and 'pathological' leads women to submit themselves in 'coerced voluntariness' to accept 'elective' cosmetic surgery in one context, and the rationalisation of sexuality and procreation in the other. Clear eugenic tendencies further commodify human life which the NRT had originally set out to serve.

Before conceptualising a desire as a right, ideology critical reflection is needed. Morgan's feminist focus is on what the pressures to be beautiful in line with market norms and successful also in one's reproductive life do to the women who submit to them. She uses Foucault's analysis of the 'docile body' to show that the operations demanded as 'rights' are a sign of power structures and alienation, by allowing oneself to be moulded by expectations and then being caught in the dynamic of the process once begun to submit to many more attempts. Her challenge strikes at the heart of the concept of 'autonomy', the crucial principle for the sake of which many other considerations – e.g. respect for human life also in its initial stages, the identity of the future child, partner's concerns, designing instead of accepting one's offspring – take second place.

Constitutional Rights Versus Economical Interests

An important factor in Morgan's biopolitical analysis of the power structures behind the beauty and fertility industries is the 'epistemic and political dominance of experts' (Morgan, 1996, p. 221). In Dietmar Mieth's portrayal of the forces struggling for the upper hand in national and international agreements on biomedical ethics, experts, biotechnological companies, and national goals of competitiveness in a globalised economy figure on one side, and the Constitution with its defence of human rights also for those who are incompetent to voice and defend them, on the other.

> One of the ethical principles which support the constitutions of legally ordered states is that some goods are not at its disposal: they may not become commodities or be commercialised. This applies, for example, to human dignity and to parts of the human body (against trade in organs or tissues). The planned or already legal inclusion of the cells of the human body (e.g. genes) among goods that can be traded is therefore rightly being disputed … The proposal that parts of the human body should be patented … which is already law in some countries, is a sign that the hopes of their use on the new biotechnological market are exerting increasing pressure on constitutional principles (Mieth, 1997, p. 120).

On the background of the German situation where the Embryo Protection Act allows *in vitro* fertilisation, but excludes freezing, experimentation, preimplantation diagnosis and selection of embryos, Mieth points out how the compromise formulae of international declarations, such as the 'European Convention on Human Rights and Biomedicine' are used by biotechnological companies as well as scientific bodies (e.g., the *Deutsche Forschungsgemeinschaft*) to challenge the high standard of legal protection with concerns for competitiveness.

> The 'location debate' – i.e. whether Europe can compete with Japan and the USA – is developing social and political pressure, and the noble argument about helping the incurably ill easily skips the questionable nature of the intermediate stages: e.g. the clearly selective methods of early genetic diagnosis. Nowadays supply has many means of strengthening itself in the guise of demand … Moreover we can doubt whether the integration of human rights in the lands of the great European Community can be secured by so many compromise formulae, which each interprets in a different way. If only the level of minimal consent is in fact affected, then this leads to pressure being put on the member states which have strict regulations to deregulate in the interest of their economic position, i.e. so that they become competitive on the market … When assisted reproduction (*in vitro* fertilization with embryo transfer), despite or because of its low success rate (14–20%), moves on to the 'market' of gene technology in

> order to lure presumably affected gene bearers into a method whose help consists in roulette with embryos and the selection of embryos, the policy often sees biotechnology as an economic factor instead of looking at it in constitutional terms. Basic goods which people need for life, like health, the inviolability of the body, dignity independent of the ability to achieve – none of this may be endangered if we want to go on living in a humane society in the future (ibid., pp. 122–3).

Against the economic pressures tilting the constitutional state towards putting financial interests first, a strong concept of basic, unalienable rights is needed. For its support, both a culture of solidarity and an understanding of law which emphasises its constitutive link to ethics are called for. Far from excluding pluralism, this link gives it a foundation, and safeguards a concept of law that is more than a pragmatic trading in rights, or simply positive law with no other argument behind it than that of authority. The ethical basis of pluralism is tolerance for different perspectives out of respect for the other. Without this normative foundation, pluralism is defenceless against the perspective of the strongest carrying the day.

A reinterpretation of a constitutive element of democracy which gives it an ethical foundation, i.e., solidarity, is already under way. True to its historical origins, it denoted society's willingness to side with its weaker members. Quite recently, this meaning has been turned upside down. Under the title of 'solidarity', these potentially weaker individuals are now pressurised to forego their claim to society's solidarity with them. In the context of genetic testing, the right not to know about one's genetic constitution becomes a duty to know; 'solidarity' now means to take preventative steps to spare society the cost of descendants who do not match the norm, a standard, moreover, which would be defined by an expert culture with no democratic legitimation. Elisabeth Hildt concludes her review of *The Right to Know and the Right not to Know* (1997) with the following outlook:

> Although in principle everybody is free to know or not to know details concerning his or her genetic make-up, it seems clear that society is highly interested in individuals who not only exert their right to know but who subsequently, by acting in solidarity, take preventative measures, especially in reproductive contexts. Bad times for a right not to know … (Hildt, 1997, p. 93).

Crisis of Values and the Significance of Ultimate Assumptions

While it is necessary to point out the democratic questionability of scientific experts such as the team who cloned Dolly and others charging ahead with unasked-for programmes of perfecting humankind and achieving a degree of 'adaptation' undreamt of in natural evolution, this critique from civil society only touches one level. Equally, it is a perceptive assessment that it indicates a conflict between the economic state and the constitutional state if industrial interests overrule fundamental human rights in order to secure market advantages on a global scale, but the problem is deeper than that. Even to declare the utter unacceptability of creating a person as a genetic blueprint of another person or to criticize that the promise of reducing human suffering is frequently reached by not allowing particular 'sufferers' to be born, remains at a level of normative ethics which can now no longer be taken for granted. It is a disquieting insight that 'before the crisis of norms lies the crisis of the intuition of values. Values, however, cannot be grounded, they can only be evoked' (Mieth, 1977, p. 6). He points to the impact of ultimate assumptions on insight into values when he adds that this 'goal in particular is the special task of a theological-ethical hermeneutics' (ibid.).

Unless the level of the values crisis is addressed, rights will continue to be eroded. The twentieth century has already witnessed the consequences of this crisis. How little resistance citizens could muster in Nazi Germany when confronted with the highest values of modern civilisation, instrumental rationality and economic efficiency, is described by the sociologist Zygmunt Bauman. The value-ladenness of technological and bureaucratic possibilities only encountered the weakness of human nature, as well as unresolved historical, national and religious antagonisms. The

> Holocaust … disclosed the weakness and fragility of human nature (of the abhorrence of murder, disinclination to violence, fear of guilty conscience and of responsibility for immoral behaviour) when confronted with the matter-of-fact efficiency of the most cherished among the products of civilization: its technology, its rational criteria of choice, its tendency to subordinate thought and action to the pragmatics of economy and effectiveness (Bauman, 1989, pp. 12–13).

My point here, clearly, is not to suggest any parallels between the genocide of the Holocaust and the new reproductive technologies but to heighten the awareness of the struggle to identify overarching values. Guiding ideals such as to make human life as controllable as possible against natural chance, e.g., include the possibility of a 'bottom-up' eugenic mentality within a pluralistic, not a totalitarian society. The battle against natural chance can only be fought at the price of instrumentalising the human object of perfection for the design imperatives of paternalising adults. No matter how good the intention, it robs the resulting person of her most original possession: her freedom which is compromised from the start by having been designed in her specificity by another human will. Instead of being regarded as a threat to be minimised, natural chance needs to be acknowledged as the guardian of freedom. The philosopher Michael Baumgartner (1987, pp. 42–3) states:

> The idea of human dignity is to be realized for the individual subject only if his beginning is not marked by the will (*Willkür*) of other human beings. Thus it can be said: only if a facticity untouched by humans and in this sense, 'fate', stands at the beginning of their own lives, can human beings determine themselves freely and that is independently of the will of other human beings. This signifies: The dignity of the human being as a being who determines himself presupposes fate. He can only be an end in himself according to the Kantian formula if what he is from his natural basis is not calculated into a context of purposes in which the human being affected can only be a means for the ends of other human beings.
>
> It may sound provocative to name this 'fate'; the point is to see that this undisposability at the beginning of human life is the condition of the truly free self-determination of a being who ... unites *animalitas* and *rationalitas* in himself ... The beginning of the human being is not marked by the will of another of his kind, but by a facticity free of will. The human being is creature, he is given to himself (*factum*) as a task (freedom), but he is not the creature of the human being. It is not the human being who stands at the beginning of human life.

Apart from the power structure it establishes, hitherto unequalled in human relationships, the mindset of technological control also loses inconspicuous, but fundamental values in its misinterpretation of the human condition. Intersubjective relationships are doomed if one merely mirrors the other, if no degree of resistance is offered to help the other go beyond her present state and become more herself. Equally, if everything around us only offers us the image of ourselves, both the challenge and the joy of the otherness of nature get lost. Even the human person turned inventor of herself might come to regret the disappearance of the resistance and the surprise of original otherness.

However, the problem and the chance of such reflections beyond the level of normative ethics is that they cannot be demonstrated by force of logic or imposed by law. If at all, they can only be accepted for the difference they make in the humane understanding of ourselves. The Oxford Christian ethicist Oliver O'Donovan evokes the values endangered by a mentality of control:

> The fate of a society which sees, wherever it looks, nothing but the products of the human will, is that it fails, when it does see some aspect of human activity which is not a matter of construction, to recognize the significance of what it sees and to think about it appropriately. This blindness in the realm of thought is the heart of what it is to be a technological culture ... Such a society is incapable of acknowledging the inappropriateness of technical intervention in certain types of activity. When every activity is understood as making, then every situation into which we act is seen as a raw material, waiting to have something made out of it. If there is no category in thought for an action which is not artifactual, then there is no restraint in action which can preserve phenomena which are not artificial. This imperils not only ... the 'environment' (as we patronisingly describe the world of things which are not human); it imperils what it is to be human, for it deprives human existence itself of certain spontaneities of being and doing, spontaneities which depend upon the reality of a world which we have not made or imagined, but which simply confronts us to evoke our love, fear, and worship (O'Donovan, 1984, p. 3).

The Impact of the New Reproductive Techniques on Theology

While philosophical ethics can be seen to be increasingly dominated by 'bioethics', the same cannot be said either for Christian ethics nor for theology as a whole. Yet, the changes in human self-understanding expressed in current biomedical developments affect the discourse of theology far beyond pastoral and social justice considerations. They touch it in more subtle ways than what could seem to be the primary provocation for theologians: the competition that arises for God as creator if the gene is declared the creative force behind our being. This response to the philosophical insight into the facticity of our existence replaces belief in a loving cause of both human and all being with an entity that only exacerbates the contingency it set out to answer. However, the real concern for theology should not be what happens to God, but what happens to God's successor if the human subject 'plays God'.

Do the new reproductive techniques mark a watershed like Kant's 'Copernican Turn' to the subject which changed the form of reflection on faith in God? Do they affect the contents of the Christian tradition, expressed in key concepts such as redemption, creation, human personhood and the equality arising from being made in the image of God? I would argue that indeed, in the age of cloning, the concepts of human nature, identity, individuality, intersubjectivity, and of life as a gift from God are under threat. If, as Oliver O'Donovan suggests, human attitudes like trust, spontaneity, wonder, awe, and gratitude, are in danger of being eroded, it is likely that with these presuppositions and correlates of faith the chances of developing an openness for God are equally dwindling.

I shall treat the theological significance of recent biomedical developments under three perspectives: human collaboration with the Creator; approaches to Christian ethics in the public realm; the compatibility of the philosophical concept of autonomy with the Christian understanding of life as a gift from God.

Human Collaboration with the Creator

It is not exaggerated to say that the divine-human relationship is affected when the creature plays the Creator. But some theologians argue that this phrase entails a 'static' concept of creation, while a 'dynamic' concept would admit that creation is never complete and that humankind, having been created with freedom and reason, owe it to their Creator to use their God-given inventiveness and creativity to 'enhance' creation. Thus, the Christian ethicist Robin Gill (1991, p. 93) judges that

> God uses human developments to further divine intentions. If the prevention of genetic deterioration is itself part of the divine intention, then those engaged in genetic research may on this understanding be seen as a part of God's work in the world.

The only cautionary note he adds is the appeal to the scientists 'to act as responsible human beings and to remember the proper ends for which they work' (ibid.).

This argumentation seems to give blanket approval to scientific research as helping 'God's work in the world'. It does not state any criteria by which to judge the 'responsibility' and 'proper ends' of the scientists. It does not specify whether the claim that 'prevention of genetic deterioration is itself part of the divine intention' is to include eugenic measures, whether it legitimates not only somatic cell therapy, but also germline therapy, and whether it stops short of cloning human beings for the benefit of other human beings. The only worry apparent in this kind of theological legitimation seems to be that theology should not once again be seen to stand in fundamental opposition to the course of scientific 'progress' which is almost automatically equated with human progress.

To my mind, this anticipating obedience to scientific development makes it more difficult to give a fair, criteria-based evaluation of the promises and the threats these projects hold. It continues the one-sided, receptive, unquestioning, and almost reverential relationship of practical theology and Christian ethics to the social and human sciences (Schweitzer, 1991).

Theologically richer, more critical and less predictable in its conclusions is the following reflection of the Free Church Federal Council and the British Council of Churches on human beings as co-creators with God:

> Such a creative, caring use of God-given powers, cannot be properly exercised without limits ... Theologically, a recognition of limits takes us right into the heart of God, who for the sake of his creation accepts the most radical limits on himself. He takes human form, shares human experience, and suffers death on the cross. Such self-limitation is part of the mystery of the divine Love (1982, p. 39).

Therefore, we cannot claim to emulate God's creativeness without also accepting the limits imposed by love for God's creatures. This qualification is especially crucial once one realises that in the case of scientific research on humans one can no longer separate basic research from its application. It is not true any more that ethics only comes in at the stage of applying results in medical therapies. In the case of humans, the difference between basic and applied research collapses (Rössler, 1987, p.164).

Thus, if we need to limit our freedom to explore and act by criteria which still need to be specified, then the question of ethical method arises.

Approaches to Christian Ethics

The problem which approach to choose is not just an academic question; it affects the public communicability of moral positions. Two dangers are obvious in applied theological ethics:

- to alienate a public in search of convincing ethical criteria by a bulwark mentality, both fundamentalist and elitist; and
- to give in to the dynamics of scientific developments, declare it a lost battle and join the bandwagon of progress with theological sanctioning.

In order to avoid these two extremes, the approach adopted and the language spoken by moral theologians should highlight the human relevance of Christian contributions to biomedical debates. Negatively, this excludes isolationist approaches and certain styles of speaking. Positively, this means that Christian ethics has something to contribute to the understanding of the shared value of autonomous freedom which I shall treat in my third point.

The moral insights put forward by churches and theologians have to be made as a contribution to the shared realm of reason. They have to be argued for. Negative examples of avoiding rational argumentation are labels such as 'intrinsically evil'. This is one of the points where the defenders of a revisionist natural law approach part company with the classical one. In her critique of the 1993 encyclical '*Veritatis Splendor*', the American moral theologian Lisa Sowle Cahill (1994, pp. 58–9) states:

> Moral evil will not be recognized as such simply because an 'intrinsic evil' flag is waved over it; substantive and reasonable arguments must be advanced about the specific nature of the acts in question.

The encyclical's claim to objectivity is also undermined by gender-exclusive language. To quote again Cahill, its

> defense of moral objectivity is ... fatally wounded by a too visible male point of view ... [it] accentuates patriarchal language and traditional gender imagery ... in Veritatis Splendor, objective human morality – the encyclical's central theme – is relentlessly advocated as necessary to 'man's' good (ibid., p. 55).

Another example of gender-exclusive language can be found in the Vatican's defence of the moral status of the embryo as a human person:

> It [the fertilized ovum] would never be made human if it were not human already ... 'The one who will be a man is already one' (Tertullian, *Apologeticum*, X, 8, quoted in Congregation of the Doctrine of the Faith, 1974).

What about the one who will be a woman? Why this resistance to translating 'homo' as 'human being'?

Apart from the gender bias, this is an argument from authority and tradition. For many, it is not clear why Tertullian should be an authority on this question. Any argument from tradition is based on shaky foundations in this area of 'mixed' ethical judgments, where theology and scientific research intersect (Mieth, 1990, p. 81). This is even more true since the tradition going back to Tertullian is contradicted by later theories of gradual ensoulment, with girl foetuses attaining the protection of ensoulment in the twelfth week of pregnancy, male embryos in the sixth week.

It is interesting to see how the Church of Scotland argues differently for the same position of attributing human rights to the embryo from fertilisation onwards:

> The moral status of the embryo and its moral claim on us do not diminish the further back we go in the stages of its development. From the moment of fertilization it has the right to be protected and treated as a human being. There is 'a serious ambiguity about an argument from the premise that the embryo is "potentially human," for the potentiality concerned is not that of becoming something else but of becoming what it essentially is' (Prof. T.F. Torrance, quoted in Kelly, 1987, p. 79).

Here, a philosophical reason is advanced. It draws on Aristotle's definition of potentiality, and can be defended as a reasonable argument: 'Anything that has within itself the principle of becoming is in potency what it will become through itself, if nothing external hinders it' (Aristotle, *Metaphysics*, Book Teta, ch. 7). It is not simply based on tradition, and it avoids the circularity looming in Tertullian's justification.

One other approach that seems to regard self-marginalisation as a theological virtue is a Christian adaptation of the group moralities advocated by Alasdair MacIntyre. It seems to hold that morality only resides in the Christian community after the catastrophe – namely, the Enlightenment – which hit morality without our noticing it, left only fragments and charred ruins, as MacIntyre's imaginative description suggests.

However, despite the controversial distinction of the good from the right, the important difference between legitimacy and legality, and a tendency towards noncommittal pluralism, liberal democratic society is not a moral vacuum. It is based on non-negotiable values, such as respect for the human person and the promotion of his freedom. The criterion we are looking for, which would limit our freedom to explore and manipulate, therefore is the mutual recognition of free human subjects as ends in themselves, not as means to other people's ends. This forbids policies which instrumentalise one human being by another. How the value of human personhood is spelt out, and whom it includes, is a matter to which Christian ethics have a decisive contribution to make.

Autonomy and Life as a Gift from God

In its original context in Kant's practical philosophy (1787), autonomy exists both within the frame of intersubjective obligation and in the parameters of human finitude. Freedom does not mean that a person has final control over her life. Autonomous freedom is aware of the fact that its existence is contingent. While it is possible to leave the question that human facticity poses open, it is also reasonable to believe in God as the ground of all being including my own as its answer.

Neither does autonomous ethics exclude the concept of God, as we can see in Kant's antinomy of practical reason where the limit questions that arise in the realisation of human freedom are admitted. The idea of God is only excluded at the level of obligation; we owe each other mutual respect because of the unconditional nature of our freedom, not because of divine command. But the Creator's power and justice are admitted at the level of meaning, when human solidarity ('virtue') and the legitimate need for happiness diverge (Kant, 1980, pp. 128–36).

There is a well-known historical observation from the critical social theorist Jürgen Habermas (1992, p. 15) on the context of origin of concepts such as freedom, person, and morality.

> Thus, I do not believe that we, as Europeans, can seriously understand concepts like morality and ethical life, person and individuality, or freedom and emancipation, without appropriating the substance of the Judeo-Christian understanding of history in terms of salvation ... without the transmission through socialisation and the transformation through philosophy of any one of the great world religions, this semantic potential could one day become inaccessible. If the remnant of the intersubjectively shared self-understanding that makes human(e) intercourse with one another possible is not to disintegrate this potential must be mastered anew by every generation.

In modernity, the validity of concepts such as morality and person clearly does not depend on religious foundations, they can stand on their own. But the question arises how long such eminent ideas can continue to shape our self-understanding if the power of reason validating them is itself battling against attempts at both genealogical and naturalising reductions. In contrast, the motivational resources to stand by the original meaning of autonomy – as self-government by reason in the context of respect for the equal rights of others – against its consumerist interpretation as 'optionalism' (Mieth, 1993, p. 275), can be renewed in religious praxis. To believe that life is a gift from God and therefore not at the disposition of humans effectively protects those fellow-human beings whose personhood is under question. It also shows 'rights' as penultimate stakes. The grateful confidence that we are gifts from God to ourselves and to each other supersedes the struggle for a right to be.

References

Almond, B. (1998), 'Gamete Transfer and Family Relationships', *Biomedical Ethics*, Vol. 3, No. 1, pp. 4–6.

Bauman, Z. (1989), *Modernity and the Holocaust*, Cambridge University Press, Cambridge.

Baumgartner, M. (1987), 'Am Anfang des menschlichen Lebens steht nicht der Mensch' in O. Marquardt and H.-J. Staudinger (eds), *Anfang und Ende des menschlichen Lebens. Medizinethische Probleme*, Vol. IV, Paderborn.

Beck, U. (1992), *Risk Society. Towards a New Modernity*, Sage, London.

Birnbacher, D. (1987), 'Gefährdet die moderne Reproduktionsmedizin die menschliche Würde?' in V. Braun, D. Mieth and D. Steigleder (eds), *Ethische und rechtliche Fragen der Gentechnologie und Reproduktionsmedizin*, J. Schweitzer, München.

Choices in Childlessness (1982), the report of a Working Party set up in July 1979 under the auspices of The Free Church Federal Council and The British Council of Churches, London.

Congregation of the Doctrine of the Faith (1974), 'Declaration on Procured Abortion (*Quaestio de abortu*)', 18 November, No. 12 in Flannery, A. (ed.) (1982), *Vatican Council Documents: More Postconciliar Documents*, Vol. 2, Dominican Publ., Dublin.

Folscheid, D. (1996), 'The Status of the Embryo from a Christian Perspective', *Studies in Christian Ethics*, 9, pp. 16–21.

Gill, R. (1991), *Christian Ethics in Secular Worlds*, T & T Clark, Edinburgh.

Habermas, J. (1992), 'Metaphysics after Kant' in *Habermas, Postmetaphysical Thinking. Philosophical Essays*, trans. by W.M. Hohengarten, MIT Press, Cambridge, Massachusetts.

Henrich, D. (1987), 'Was ist Metaphysik – was Moderne? Zwölf Thesen gegen Jürgen Habermas' in D. Henrich, *Konzepte*, Suhrkamp, Frankfurt.

Hildt, E. (1997), review of *The Right to Know and the Right not to Know* [R. Chadwick, M. Levitt and D. Shickle (eds), Avebury: Aldershot, 1997], *Biomedical Ethics*, Vol. 2, No. 3, pp. 92–3.

Honnefelder, L. (1993), 'Der Streit um die Person in der Ethik', *Philosophisches Jahrbuch*, 100, pp. 246–65.

Honnefelder, L. (1996a), 'Person und Menschenwürde' in L. Honnefelder and G. Krieger (eds), *Philosophische Propädeutik, Bd. 2: Ethik*, UTB, Paderborn.

Honnefelder, L. (1996b), 'Nature and Status of the Embryo: Philosophical Aspects', unpublished manuscript for the Council of Europe's Third Symposion of Bioethics, *Medically Assisted Procreation and the Protection of the Human Embryo*, 15–18 December, Strasbourg.

Junker-Kenny, M. (1998a), 'The Moral Status of the Embryo' in M. Junker-Kenny and L. Sowle Cahill (eds), *The Ethics of Genetic Engineering*, *Concilium 1998/2*, SCM Press, London.

Junker-Kenny, M. (1998b), 'Der moralische Status des Embryos im Kontext der Reproduktionsmedizin' in M. Düwell and D. Mieth (eds.), *Ethik in der Humangenetik*, Francke, Tübingen.

Junker-Kenny, M. (1999), 'Categorical Arguments: Pro life versus pro choice?' in E. Hildt and S. Graumann (eds.), *Genetics in Human Reproduction*, Ashgate Publishing Ltd, Aldershot.

Kant, I. (1980), *Critique of Practical Reason* (1787), trans. by L.W. Beck, Bobbs-Merrill, Indianapolis.

Kelly, K.T. (1987), *Life and Love. Towards a Christian Dialogue on Bioethical Questions*, Collins, London.

Kummer, C., SJ (1997), 'Extrauterine Abtreibung? Sachargumente für eine Bestimmung des embryonalen Lebensbeginns', *Stimmen der Zeit*, 215, pp. 11–16.

Mieth, D. (1977), *Moral und Erfahrung. Beiträge zur theologisch-ethischen Hermeneutik*, Universitätsverlag, Freiburg/Schweiz.

Mieth, D. (1990), *Geburtenregelung. Ein Konflikt in der katholischen Kirche*, Grünewald, Mainz.

Mieth, D. (1993), 'The Problem of "Justified Interests" in Genome Analysis. A Socioethical Approach' in H. Haker et al. (eds), *The Ethics of Human Genome Analysis. European Perspectives*, Attempto Verlag, Tübingen.

Mieth, D. (1997), 'The Market and the Inviolability of Human Dignity in the Perspective of Biomedical Technology' in D. Mieth and M. Vidal (eds), *Outside the Market No Salvation, Concilium 1997/2*, SCM Press, London.

Morgan, K.P. (1997), 'Gender Rites and Rights: The Biopolitics of Beauty and Fertility' in L.W. Sumner and J. Boyle (eds) (1997), *Philosophical Perspectives on Bioethics*, University of Toronto Press, Toronto/Marston, Abingdon.

O'Donovan, O. (1984), *Begotten or Made?*, Clarendon Press, Oxford.

Pröpper, T. (1995), 'Autonomie und Solidarität. Begründungsprobleme sozialethischer Verpflichtung' in E. Arens (ed.), *Anerkennung der Anderen*, Herder, Freiburg.

Ricoeur, P. (1992), *Oneself as Another*, University of Chicago Press, Chicago.

Rössler, D. (1987), 'Was verändert sich in der Ethik?' in V. Braun et al. (eds), *Ethische und rechtliche Fragen der Gentechnologie und der Reproduktionsmedizin*, J. Schweitzer Verlag, München.

Schweitzer, F. (1991), 'Praktische Theologie, Kultur der Gegenwart und die Sozialwissenschaften – Interdisziplinäre Beziehungen und die Einheit der Disziplin' in K.-H. Nipkow et al. (eds), *Praktische Theologie und Kultur der Gegenwart*, Gütersloher Verlagshaus, Gütersloh.

Sowle Cahill, L. (1994), 'Accent on the Masculine' in J. Wilkins (ed.), *Understanding Veritatis Splendor*, SPCK, London.

Steigleder, K. (1993), 'Menschenwürde – Zu den Orientierungsleistungen eines Fundamentalbegriffs normativer Ethik' in J.P. Wils (ed.) (1993), *Orientierung durch Ethik? Eine Zwischenbilanz*, Schöningh, Paderborn.

Steigleder, K. (1997), 'The Moral Significance of Potential Persons' in E. Hildt and D. Mieth (eds), In Vitro *Fertilization in the 1990s*, Ashgate Publishing Ltd, Aldershot.

Wils, J.-P. (ed.) (1993), *Orientierung durch Ethik? Eine Zwischenbilanz*, Schöningh, Paderborn.

Wimmer, R. (1995), 'Anthropologie und Ethik. Erkundungen in unübersichtlichem Gelände' in C. Demmerling et al. (eds), *Vernunft und Lebenspraxis*, Suhrkamp, Frankfurt.

Wimmer, R. (1997), 'Ethical Aspects of the Concept of Person', *Biomedical Ethics*, 2, pp. 86–9.

Wimmer, R. (1998), 'Bioethical Aspects of the Concept of Person', unpublished manuscript, partly to be published in E. Hildt and S. Graumann (eds.) (1999), *Genetics in Human Reproduction*, Ashgate Publishing Ltd, Aldershot.

Contributors

Dan Bradley is a lecturer in the Genetics Department at Trinity College Dublin. He leads a research group which is investigating the evolution of the genomes of domestic animals, particularly cattle. He also has an interest in human evolution, especially in the origins of European populations.

Neville Cox is a lecturer in European Law at Trinity College Dublin. He lectures in comparative law, legal systems and methods, medicine and the law and has published on Irish blasphemy law.

Sheila Greene is a Senior Lecturer in Psychology and the co-founder and co-academic director of The Children's Research Centre at Trinity College Dublin. She is a Fellow of the Psychological Society of Ireland and a Fellow of Trinity College Dublin. Her recent research and publications are in developmental psychology and the psychology of women.

Hille Haker is research assistant at the Department of Theological Ethics at the Faculty for Catholic Theology and member of the Center for Ethics in the Sciences and Humanities, University of Tübingen, Germany. Since 1998, she has been coordinator of the European Network for Biomedical Ethics, a concerted action funded by the European Commission from 1996–99. She is also a member of the German Academy of Ethics in Medicine and member of the editorial board of the journal *Ethical Perspectives*. She is co-editor of *Ethics of Human Genome Analysis. European Perspectives*, Attempto Verlag, Tübingen, 1993. Her main research interests are foundational aspects of bioethics and medical ethics, genetic diagnosis and genetic counselling, as well as moral personhood and identity.

Maureen Junker-Kenny is at present Head of the School of Hebrew, Biblical and Theological Studies, Trinity College Dublin where she lectures in Christian Ethics/Practical Theology. She is a member of the directorial board of the international Catholic journal *Concilium* and co-editor of the International *Journal of Practical Theology*. She is a Fellow of Trinity College Dublin. Her research interests are in discourse ethics and Christian praxis, biomedical ethics, as well as conditions for faith in late modernity.

Helga Kuhlmann is Professor of Protestant Theology with a focus on Systematic Theology and Ecumenics at the University of Paderborn, Germany. Her publications include *Die theologische Ethik Albrecht Ritschls*, Munich, 1992; editor of *'Und drinnen waltet die züchtige Hausfrau'. Zur Ethik der Geschlechterdifferenz*, Gütersloh, 1995, as well as articles in systematic and feminist theology.

David McConnell, M.R.I.A., F.Z.S.I. is Professor of Genetics at Trinity College Dublin. He is a molecular geneticist, educated at Trinity College (BA, 1966) and the California Institute of Technology (PhD, 1971). He was an Eleanor Roosevelt Fellow of the IUAC at Harvard University (1976–77), he is a member of the European Molecular Biology Organisation, a member of the Irish Council for Science, Technology and Innovation and President of the Adelaide Hospital Society. He is a Fellow of Trinity College Dublin.

Dietmar Mieth is Professor of Theological Ethics at the Faculty for Catholic Theology at the University of Tübingen and Head of the interdisciplinary Center for Ethics in the Sciences and Humanities, Tübingen, Germany. Since 1994, he has been a member of the advisory group to the European Commission, 'Group of Advisers for Ethical Implications for Biotechnology', since 1998 a member of its successor committee, 'European Group of Ethics'. From 1996 onwards he has been the Head of the European Network for Biomedical Ethics. His areas of specialisation are social ethics, business ethics, medical and bioethics, religious experience.